M000287115

International Launch Site Guide

International Launch Site Guide
Second Edition

Steven R. Strom

The Aerospace Press • El Segundo, California

American Institute of Aeronautics and Astronautics, Inc. • Reston, Virginia

The Aerospace Press
2350 E. El Segundo Boulevard
El Segundo, California 90245-4691

American Institute of Aeronautics and Astronautics, Inc.
1801 Alexander Bell Drive
Reston, Virginia 20191-4344

Library of Congress Cataloging-in-Publication Data

Strom, Steven R.
  International launch site guide / Steven R. Strom.-- 2nd ed.
     p. cm.
  Rev. ed. of: International launch site guide. 1994.
  Includes index.
  ISBN 1-884989-16-0
  1. Launch complexes (Astronautics)--Directories. I. International launch
site guide. II. Title.

  TL4020.I58 2005
  629.47'8--dc22
                              2005009754

On the cover: Titan IV/Milstar payload at U.S. Eastern Range
Photo by Russ Underwood/© 2005 Lockheed Martin
Cover design by Karl Jacobs

Data and information appearing in this book are for informational purposes only.
The publishers and the authors are not responsible for any injury or damage resulting from use or reliance, nor do the publishers or the authors warrant that use or
reliance will be free from privately owned rights.

The material in this book was reviewed by the Air Force Space and Missile Systems Center, and approved for public release.

Contents

Preface vii

Introduction ix

Acknowledgments xiii

1 Australia 1
Woomera Rocket Range 1

2 Brazil 9
Alcântara Launch Center 11
Barreira do Inferno Sounding Rocket Range 15

3 Canada 21
Churchill Rocket & Research Range 21

4 China 27
Jiuquan Satellite Launch Center 28
Xichang Satellite Launch Center 31
Taiyuan Satellite Launch Center 34

5 France and French Guiana 39
Guiana Space Center 39

6 India 47
Satish Dhawan Space Centre 47

7 Israel 53
Palmachim Launch Site 53

8 Japan 57
Uchinoura Space Center 60
Tanegashima Space Center 63

9 Norway 69
Andøya Rocket Range 69

10 Pakistan 73
SUPARCO Flight Test Range 73

Contents

11 Russia and Kazakhstan 77
Plesetsk Cosmodrome 79
Baikonur Cosmodrome 85

12 Sweden 95
Esrange 95

13 United States. 101
Eastern Range 105
Western Range 117
Wallops Flight Facility 130

14 A Multinational Site. 139
Sea Launch 139

Appendix I: Proposed Commercial U.S. Spaceports 147

Appendix II: Points of Contact Quick Reference. 157

Index 165

Preface

The second edition of the *International Launch Site Guide* is an updated and revised version of the first edition, which was edited by Roy M. Chiulli and published by The Aerospace Press in 1994. The book is a general guide to the world's current principal operating launch sites as well as some sites that may be operational in the near future. This edition incorporates changes from the past decade within both the space industry and national space programs, some of which reflect the new geopolitical realities of the post–Cold War world. One site that has been added, Sea Launch, particularly reflects these changes, because the multinational partnership that operates the site—a team that includes the United States and Ukraine—was foreseen by only a few visionaries in the early 1990s.

While the original *Guide* was written to provide information on sites capable of launching commercial payloads, the second edition also includes sites for other reasons: because they are important to scientific studies (e.g., Sweden), because they have strategic or political importance (Israel), or because they are new (Sea Launch). Some sites from the original, such as Cape San Blas and the San Marco Equatorial Range, have been excluded because they have been inactive or deactivated since 1994. As with the first edition, only selected sites are included here, because of the large number of launch sites in the world. The inclusion of a site, however, is largely reflective of its importance in the international space community.

Not only has the selection of sites been changed, but the chapters themselves have also been updated. Each chapter, devoted to a country (or, in some cases, to related countries), includes expanded overview material that places the country's launch sites in historical and geographical perspective. Like the first edition, this one discusses each site's operational capabilities.

The information contained in the second edition was obtained from publicly available sources that include both printed materials and World Wide Web sites. The Aerospace Press intends the *International Launch Site Guide* for reference purposes only, not for specific technical applications. Please consult a launch site directly to obtain the most up-to-date information on that site. The author, editor, and publishers have used their best efforts to ensure the accuracy of the book's contents but assume no liability for errors or omissions.

Early photo of the Western Range at Vandenberg Air Force Base.

Introduction

Since its founding in 1960, The Aerospace Corporation has been involved in many aspects of the U.S. space program, in both the military and civilian sectors. Aerospace was created as a nonprofit corporation to provide technical assistance to the Department of Defense for space activities associated with national security. The company's work and research for many of the nation's principal launch vehicles led to a corollary involvement with U.S. launch sites, a relationship that extends to the present. Throughout the past forty-five years, the company has been a continuous participant in numerous activities at both the Western Test Range (Vandenberg Air Force Base in California) and the Eastern Test Range (Cape Canaveral Air Force Station in Florida). Recently Aerospace has also provided assistance and technical expertise to a number of international agencies. So it is from a logical outgrowth of corporate interest and relationships that The Aerospace Press, the corporation's publishing arm, now introduces the second edition of the *International Launch Site Guide*.

The Aerospace Corporation began operations by assisting the Air Force with general systems engineering and technical direction (GSE/TD) for the launch vehicles used in two of the most historically important crewed U.S. space programs, Project Mercury and the Gemini program, during the formative years of space exploration. For both of these programs, Aerospace developed and implemented a Pilot Safety Program and a system of man-rating the Atlas and Titan rockets that were used as launch vehicles, thus guaranteeing the astronauts' safety and the programs' ultimate success. Aerospace also devised the Titan Integrate-Transfer-Launch (ITL) system during a period when the company provided GSE/TD for the design of the Titan facilities, the launch equipment, and the ground-systems handling apparatus. During these early years of the space program, Aerospace engineers, working alongside the Army's Corps of Engineers and NASA, had input in the planning and design of launchpads at Cape Canaveral. Aerospace helped oversee the construction of the ITL complex at the Cape, which included launch complexes 40 and 41, and the company supplied GSE/TD for the launches of Titan IIIA's from Launch Complex 20.

In addition to working with these early NASA programs, Aerospace was concurrently assisting the Air Force with the development of military crewed space programs. One of the first program offices formed at Aerospace was the Dyna-Soar program office, established in the summer of 1960. As with the Mercury and Gemini programs, Aerospace provided GSE/TD for the visionary Dyna-Soar program, cancelled in 1963. (The Dyna-Soar would have been the world's first reusable, orbital space plane. Although it never flew, research for the Dyna-Soar was later applied to the space shuttle and other U.S. space systems.) Beginning in the mid-1960s, Aerospace became a vital participant in the Air Force's Manned Orbiting Laboratory (MOL) program, which eventually became the corporation's largest program office. One of the largest and most complex projects that Aerospace ever

Introduction

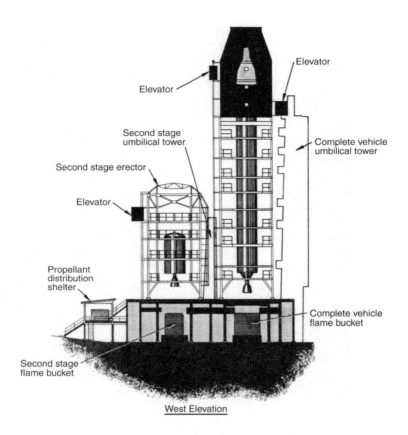

Elevator

Elevator

Second stage umbilical tower

Complete vehicle umbilical tower

Second stage erector

Elevator

Propellant distribution shelter

Complete vehicle flame bucket

Second stage flame bucket

West Elevation

AEROSPACE CORPORATION
Atlantic Missile Range Office

MISSILE ERECTORS
(COMPLEX 19 / GEMINI)
CAPE CANAVERAL, FLORIDA

AS7-3B-62

| REQUESTOR: H.E.ELEY | | CHECKED BY: J.L.GREGORY |
| DRAWN BY: G.M.STAMATAKIS | REVISED BY: | DATE: SEPT. 21, 1962 |

This elevation drawing of a Gemini launch configuration depicts a Gemini capsule atop a Titan II rocket at Cape Canaveral's Launch Complex 19. The complex supported 12 Gemini missions (two of them unmanned) during 1964–1966. This engineering drawing was delineated in 1962 for The Aerospace Corporation's Atlantic Missile Range Office. Launch Complex 19 was declared a national historic landmark in 1984.

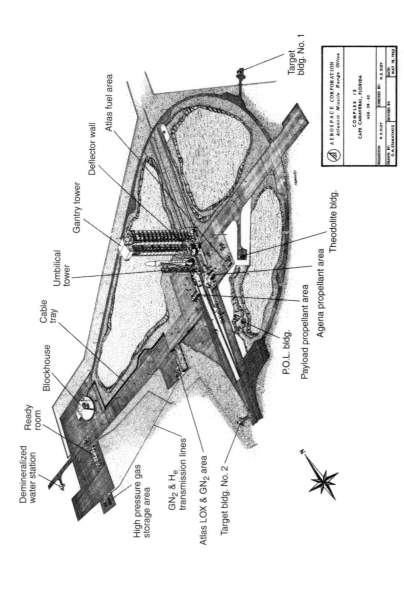

Demineralized water station

Ready room

Blockhouse

Cable tray

Umbilical tower

Gantry tower

Deflector wall

Atlas fuel area

High pressure gas storage area

GN₂ & He transmission lines

Atlas LOX & GN₂ area

Target bldg. No. 2

P.O.L. bldg.

Payload propellant area

Agena propellant area

Theodolite bldg.

Target bldg. No. 1

N

AEROSPACE CORPORATION
Atlantic Missile Range Office

COMPLEX 13
CAPE CANAVERAL, FLORIDA
A56-29-62

REQUESTOR: H.E.RILEY	CHECKED BY: H.E.RILEY
DRAWN BY:	REVISED BY:	DATE:
G.N.STAMATAKIS		MAY 18, 1962

This 1962 aerial perspective drawing shows the layout of Launch Complex 13 at Cape Canaveral. This complex supported launches of Atlas and Atlas/Agena launch vehicles from 1958 to 1978, when it was deactivated. Personnel from The Aerospace Corporation had input into the modification process of several Cape Canaveral launch complexes.

assisted with was the construction of Space Launch Complex 6 (SLC-6) at Vandenberg. SLC-6 was intended as the launchpad for the MOL launch vehicle, the Titan IIIM. Although MOL was cancelled in 1969, the program proved to be an invaluable testing ground for some important early space concepts and technologies.

Following the cancellation of MOL and the conclusion of the Apollo program, by the early 1970s Aerospace was performing advanced program analysis and studies for the space shuttle project. Aerospace worked closely with the Air Force, which was developing its own shuttle launch capacities at Vandenberg. Beginning in 1979, some $4 billion was spent to modify the deactivated SLC-6 complex in preparation for shuttle launches. After the space shuttle successfully flew in 1981, Aerospace opened an office at the Kennedy Space Center (KSC) to assist the transition of shuttle technology from NASA to the Air Force, although the Aerospace KSC team was not involved in shuttle launch operations.

Following the explosion of the space shuttle *Challenger* in January 1986, the Air Force cancelled its shuttle operations at Vandenberg and reactivated expendable launch vehicle production. Once the decision to do this was made, Aerospace personnel provided the Air Force with valuable assistance as a result of their many years of experience with heritage launch vehicles, including Atlas and Titan. When work began on Evolved Expendable Launch Vehicles (EELVs) in the 1990s, Aerospace became a valuable participant in the development of these new launch vehicles. At the beginning of the twenty-first century, Aerospace played an important role in the first successful launches of both the Atlas V and the Delta IV EELVs. By 2005, the forty-fifth anniversary of Aerospace, the company had been involved with virtually every major component of the U.S. space program. It continues to play a major role in the development and planning of future space systems.

Acknowledgments

Special thanks to Jon S. Bach for his skillful editing throughout the project and to Rick Mortimer and Rick Francis for reviewing the manuscript; to Roy M. Chiulli for producing the original manuscript on which this work is based; and to Donna J. Born and David J. Evans of The Aerospace Institute for their early and continued support. Thanks are also in order to artists Karl Jacobs (cover design) and John Hoyem (illustrations) for their high-quality work. Thanks to Jimmy Kephart and Pete Portanova for research assistance, and to Joyce Hardy of The Aerospace Library for her support.

Source Notes

Photographs

Chapter 1: Launch of the University of Queensland's HyShot rocket from the Woomera Range in 2002. Used courtesy of University of Queensland.

Chapter 3: An Aerobee-Hi research rocket being fired at the IGY(International Geophysical Year) rocket launching site in Fort Churchill, Manitoba, Canada. Used courtesy of U.S. National Academy of Sciences.

Chapter 5: *Ariane 5* at the Launch Zone no. 3 at CSG (Centre Spatial Guyanais). Used courtesy of CNES (Centre National d'Études Spatial).

Chapter 8: *H-IIA* launch vehicle GTV, ground firing test at Osaki Range, Yoshinobu Launch Complex. Used courtesy of JAXA (Japan Aerospace Exploration Agency).

Chapter 11: Proton rocket on launchpad at Baikonur Cosmodrome. Used courtesy of Lockheed Martin Space Systems.

Chapter 13: Liftoff of STS-58 from pad 39B; aerial view of Launch Complex 39. Both used courtesy of NASA (National Aeronautics and Space Administration).

Chapter 14: Transfer of the *Zenit-3SL* launch vehicle from the ship to the launch platform; SL GXIII liftoff Galaxy XIII/Horizons 1; launch of Loral's Telstar 14/ Estrela do Sul 1. All used courtesy of Sea Launch.

Maps

Maps of the following launch sites have been used courtesy of Mark Wade: Woomera Rocket Range (Chapter 1); Alcântara Launch Center (Chapter 2); Jiuquan Satellite Launch Center, Xichang Satellite Launch Center, and Taiyuan Satellite Launch Center (Chapter 4); Guiana Space Center (Chapter 5); Palmachim Launch Site (Chapter 7); Uchinoura Space Center and Tanegashima Space Center (Chapter 8); SUPARCO flight test range at Sonmiani (Chapter 10); Plesetsk Cosmodrome and Baikonur Cosmodrome (Chapter 11); U.S. Eastern Range, U.S. Western Range, and Wallops Flight Facility (Chapter 13). These site maps are all © Mark Wade.

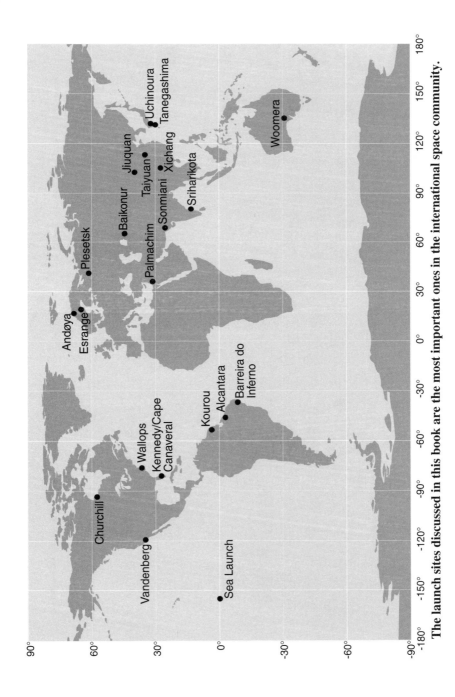

The launch sites discussed in this book are the most important ones in the international space community.

1 Australia

Woomera Rocket Range
• orbital • 31.0 deg south • 136.5 deg east

Overview

The Woomera Rocket Range is one of the world's largest launch sites. Officially known as the Woomera Prohibited Area (WPA), the facility covers some 127,000 sq km, which is almost exactly the size of England. Since Woomera's first missile was launched in 1949, the range has played an important role in the history of rocketry. It has been the site of guided weapons testing, space capsule launch and retrieval, sounding rocket tests, satellite launches, and tracking of early spacecraft, including the crewed spacecraft of the American Mercury program.

In broad terms, the establishment of the spaceport in 1947 marked the beginning of a joint Australia–United Kingdom agreement to develop and test guided weapons systems. More specifically, the British needed access to a launch site within the Commonwealth to test-fire the large number of missile systems planned in the postwar era. Several potential locations were identified, and the vast, sparsely populated, flat terrain of the state of South Australia was selected.

Woomera, an Australian aboriginal word meaning "spear thrower," refers not only to the WPA but also to Woomera township, which contains support facilities for the rocket range. The town of Woomera is located in the southeastern corner of the WPA, about 500 km northwest of Adelaide, the capital of South Australia. A major facet of the WPA is the primary trials area known as the Woomera Instrumented Range (WIR), located some 40 km from Woomera township. Available launch vehicle sites for customers are located within the WIR.

During the 1950s and 1960s, as the Cold War between the Soviet Union and the United States and its Western allies intensified, the WPA acquired an increasingly important role in Western defense-related development work. The range served as the launch site for numerous rockets, including Skylark, Black Knight, and Europa. The formation of the European Launcher Development Organization (ELDO) in 1962 was another important factor in the increase of WPA activities, because ELDO member states used the range for a large number of rocket tests for the remainder of the decade. In 1967, Australia achieved the distinction of launching its own satellite into orbit, using an American Redstone rocket as the launch vehicle. During this period, the population of the town of Woomera roughly doubled from approximately 3000 people in 1950 to more than 6000 in the 1960s. Great Britain launched its first satellite, Prospero, from Woomera in 1971, but cutbacks in British defense programs soon reduced the scale of Woomera's launch activities, and the British ceased operations there in 1976.

Although rocket testing and launches at the WPA reached a peak in the 1960s, Woomera took on an additional role as support center for the Nurrungar tracking station, established in 1970 to assist the U.S. Defense Support Program (DSP). The Joint Defense Facility Nurrungar, some 20 km from the town of Woomera, was a joint operation of the Australia Defence Forces and the U.S. Air Force Space Command, and its original purpose was to assist the DSP with the detection

1

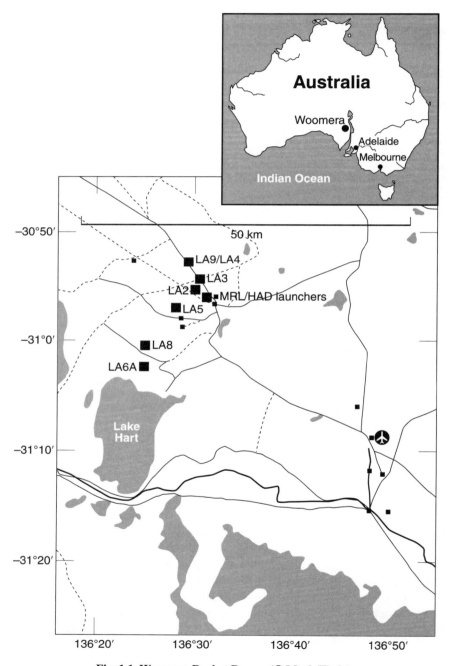

Fig. 1.1. Woomera Rocket Range. (© Mark Wade)

and tracking of Russian missile launches and nuclear detonations. Operations at Nurrungar provided important assistance during the Persian Gulf War in 1991 but subsequently were gradually phased out and the facility was closed in 1999.

After the 1970s, use of the Woomera range declined, and the last major series of launches was a number of Black Brant sounding rockets launched for NASA in 1995. Today, the population of the WPA numbers only about 1300 permanent residents. Several recent, highly publicized projects to reactivate the range ultimately collapsed, most notably a project proposed by the SpaceLift Australia company to provide launch services for satellite payloads from Woomera. In March 2005, the Kistler Aerospace Corporation announced plans to use the WPA as a launch site for its planned K-1 reusable launch vehicle. Kistler intends to attempt its first launch from the WPA in 2007. Australia signed an agreement with Japan in 2001 to allow the Japanese National Aerospace Laboratory (NAL) to launch a series of tests of NAL-developed, rocket-powered experimental planes. The first test of the planned supersonic plane ended disastrously when the test model exploded shortly after liftoff in July 2002. The Australian Space Council has recently considered the construction of new launchpads on the WPA to attract continued use of the range.

Launch Site Description

The WPA is currently administered by the Australian Department of Defence, which allows it to be used for not only military but also approved civil and commercial purposes. The Area Administrator Woomera is in charge of the Defence Support Centre Woomera (DSCW), which has responsibility for WPA management and operations.

In addition to its largely flat, featureless, and vast terrain, which allows easier access for test object recovery and provides an important safety valve for hazardous launch activities, the WPA's desert climate is another positive feature of the range. Rainfall is rare, which results in few overcast days, and the climate is generally warm and dry, with only a brief, generally mild winter. The area's stable climatic conditions virtually assure the ability to conduct year-round operations, with little downtime resulting from poor weather conditions. An important aspect of the WPA's remote location is the ability of the range to enforce strict security when required and to limit overflights by air traffic during launch and test activities. The low population density of the WPA is another range attribute.

The Woomera Instrumented Range (WIR), an area of approximately 2000 sq km, includes the WPA's launch vehicle facilities. The WIR is under the jurisdiction of the Aircraft Research and Development Unit of the Royal Australian Air Force. The Woomera Rangehead is located in the southeast portion of the WIR, and its primary facilities include three launch sites for missiles and sounding rockets with launch control facilities, assembly and checkout facilities for flight vehicles, an instrumentation building for managing range activities and safety, and a meteorological shed for the release of weather balloons.

Additional Rangehead facilities and specifications include:

- fully prepared and surveyed sites
- security protection
- rocket motor and explosive ordnance storage facilities
- helicopter landing pad and small airfield
- optical instrumentation
- tracking data system
- local WIR communications system
- meteorological facilities
- large launch vehicle and payload test shop/assembly facilities
- hardware recovery facilities

The three operational launch areas currently located within the WIR are referred to as LA1, LA2, and LA9. LA1, recently refurbished, now contains a Japanese-owned launcher. LA2 has two sounding rocket launchers and has been the Rangehead's principal launch site for many years. LA9 also serves as a site for launching sounding rockets.

The WPA also includes the Lake Hart Air Weapons Range (LHAWR) on the northern shore of Lake Hart, a dry salt lake. This range is used for the release of live munitions drops from aircraft and ground-based missile and artillery firings. The LHAWR also contains the former launch areas LA6a and LA6b, originally constructed for Blue Streak missile tests. Two large, deep concrete launch pedestals remain here and have the potential to be refurbished and made operational for large-launch-vehicle programs.

Payload Accommodations and Support Facilities

Payloads and launch vehicles can be transported to Woomera township using air, rail, or road transport. Port facilities are available at Adelaide, which is connected to Woomera by highway. Transport from Woomera to the launching area (WIR) is by road. Test shop integration and assembly facilities are available at Woomera township or the Rangehead.

Only eight km from the town of Woomera, a large airport with a 2375 m runway is available for use by both civil and military airplanes. It can handle large freight aircraft, including the Lockheed C-5A and the Boeing 747, and can also support aircraft that specialize in the air launch of space launch vehicles. The airport also has three large hangars with some 9000 sq m of space.

The town of Woomera offers full facilities and services in support of range activities, including accommodations for personnel utilizing the WPA, recreational facilities, medical care, and communications services.

Fig. 1.2. Launch of the University of Queensland's HyShot rocket from the Woomera Range in 2002. (Courtesy University of Queensland)

Acronyms

DSCW	Defence Support Center Woomera
DSP	Defense Support Program (U.S.A.)
ELDO	European Launcher Development Organization
LHAWR	Lake Hart Air Weapons Range
NAL	National Aerospace Laboratory (Japan)
NASA	National Aeronautics and Space Administration (U.S.A.)
WIR	Woomera Instrumented Range
WPA	Woomera Prohibited Area

Points of Contact

Area Administrator Woomera
Defence Support Centre Woomera
Dewrang Avenue
Woomera, South Australia 5720
Australia
Voice: +61 8 8674 3201

Fax: +61 8 8674 3308
E-mail: aaw@dfence.gov.au

Commander ARDU
Aircraft Research and Development Unit
RAAF Base Edinburgh
Edinburgh, South Australia 5111
Australia
Voice: +61 8 8393 2111
Fax: +61 8 8393 2498

Space Policy Unit
Department of Industry, Science and Resources
P.O. Box 9839
Canberra, Australian Capital Territory 260
Australia
Voice: +61 6 213 7246
Fax: +61 6 213 7249
Telex: AA62654
E-mail: egrohovaz@isr.gov.au

References

"Australia in Space," *AeroSpaceGuide.net*, 2005,
<http://www.aerospaceguide.net/worldspace/australia_in_space.html>
(17 January 2005).

BAE Systems and Department of Defence (Australia), "Range," *Woomera South Australia*, <http://www.woomerasa.com.au/PAGE.CFM/DC5AF086-9921-4C67-BAF1-D439657F9BFC> (17 January 2005).

Defence South Australia, <http://www.defence-sa.com/html/infrastructure/woomera.htm> (17 January 2005)

Flinders Ranges Research, <http://www.southaustralianhistory.com.au/woomera.htm> (inactive).

M. King, ed., *The Lowdown: Australian Space Development in Focus*,
<http://www.lowdown.com.au/> (1 March 2005).

P. Morton, *Fire Across the Desert: Woomera and the Anglo-Australian Joint Project, 1946–1980* (Australian Government Publishing Service, Canberra, 1989).

M. T. Rigby, *Woomera on the Web*, 2005, <http://homepage.powerup.com.au/~woomera/> (17 January 2005).

M. Wade, ed., "Woomera," *Encyclopedia Astronautica,* 2003, <http://www.astronautix.com/sites/woomera.htm> (17 January 2005).

2 Brazil

Overview

The Brazilian space program can trace its beginnings to the formative years of the space age. Like other people around the world, Brazilians were excited in the late 1950s by the launches of *Sputnik 1* and other early satellites and hoped their nation would participate in the exploration of space. Since that time, the Brazilian government has come to see space science and technology as vital components of its efforts to develop the national economy.

Brazil has a long history of aviation achievements, and air travel was recognized from its beginnings as a necessity for transportation within such a huge country. With Brazil's aviation background and its geographical location, which allows geosynchronous launches from its equatorial regions, it was natural for many Brazilians to eventually turn their interests toward space exploration.

In 1961, President Jânio Quadros authorized the creation of the Organizing Group for the National Committee on Space Activities, which by 1971 had evolved into the National Institute for Space Research (Instituto Nacional de Pesquisas Espaciais, or INPE). By the mid-1960s, Brazil was already launching sounding rockets. The Brazilian government codified its national space aspirations in 1979 with the creation of the Complete Brazilian Space Mission (Missão Espacial Completa Brasileira, or MECB) and the announcement of a comprehensive program that included Brazil's long-term space goals. These objectives included the development and manufacture of a rocket, the establishment of a launch site on Brazilian territory, and the design and manufacture of a satellite.

In 1994, as a sign of the Brazilian government's recognition of the importance of the space sector, it created the Brazilian Space Agency. This civilian agency, which reports to the Office of the President of the Republic, has responsibility for formulating the nation's space policy. Brazil became a participant in the International Space Station (ISS) program after an agreement was signed in 1997 with the United States that allowed for the training of Brazilian astronaut Major Marcos Pontes at the Johnson Space Center in Houston. Pontes completed his training and is currently waiting for an opportunity to fly onboard the ISS. He will be the first Brazilian to fly in space.

The Brazilian commitment to space research and development has been demonstrated by the construction of two launch sites—the Barreira do Inferno, which was developed in 1964 and is intended primarily for suborbital launches, and the more ambitious Alcântara Launch Center. Alcântara opened in 1990 as the site for the launch of Brazil's indigenous Veículo Lançador de Satelites (VLS) rocket—a four-stage, solid-propellant rocket designed to launch a 200 kg satellite into a 500 km circular orbit. So far, attempts to launch the VLS (in 1997, 1999, and 2003) have been unsuccessful. Nevertheless Brazil hopes that Alcântara will one day become a major international site for the launching of commercial satellites.

Brazil took an important step toward the goals of its space program in November 2004, when President Luis Inácio Lula da Silva signed an agreement with Russian President Vladimir Putin to cooperate on a number of space projects. The memorandum of understanding called for joint utilization of the Alcântara

Fig. 2.1. Location of the two Brazilian launch sites.

launch site, the development and launch of geostationary satellites, and mutual cooperation on the development of the VLS, as well as new models of Brazilian launch vehicles. The signing of the agreement clearly demonstrated the Brazilian government's commitment to become a major player in the arena of space exploration and the commercial utilization of space.

Alcântara Launch Center
• orbital • 2.28 deg south • 44.38 deg west

Launch Site Description

The Alcântara Launch Center, closer to the equator than any other land-based launch site in the world, was developed for the launching of the VLS, Brazil's first orbital launch vehicle. Alcântara is located near the town of São Luis, in the state of Maranhão, along Brazil's northeastern Atlantic coast. In addition to proximity to the equator—it is less than 3 deg of latitude away, which allows for lower launch costs—this spot was chosen because of its low population density and its coastal location, which enables launches to take place over water, potentially limiting damage in case of launch failure. Launch azimuths range from 10 to 100 deg. Development of the 620 sq km site began in 1983, and it was formally opened in February 1990. To date, Brazil has invested approximately U.S.$300 million into developing the site. The Brazilian government has promoted Alcântara as an alternative to the Guiana Space Center in neighboring French Guiana for launching satellites into geostationary orbit.

The weather at Alcântara includes a wet season and a dry one, with the calendar year neatly split in half. The months from January to June have the greatest rainfall, with the rain peaking during April and May. The remainder of the year is usually quite dry, with some months having drought-like conditions.

Launches from Alcântara

The first two attempts to launch the VLS-1 from Alcântara, in 1997 and 1999, both failed and resulted in the destruction of the launch vehicles. A third effort ended catastrophically in August 2003, when an electrical flaw triggered one of the rocket's four solid-fuel boosters three days before the scheduled launch. The resulting explosion, which killed 21 people, was a terrible setback for the Brazilian space program. A government report issued in March 2004 blamed the accident on poor management practices and a lack of proper funding for the space program. Current plans call for another VLS launch attempt in 2006.

Apart from these three attempts at orbital launches, Alcântara has been the site of numerous launchings of sounding rockets over the years, including the Brazilian *Sonda-1* and *Sonda-2*, with nearly 300 launches taking place since 1989. Brazil has also advocated the site's use by other nations, which could become an additional source of revenue. China has suggested the use of Alcântara as a launch site for its Long March rockets, and the Ukraine signed an agreement with Brazil in 2003 to launch its *Cyclone-4* rockets from Alcântara. In addition, Israel agreed in 2003 to launch some of its Shavit rockets from the site, although launch dates have not yet been determined. The possibility of commercial operations at the site has also been discussed with several private companies, because savings of up to 30 percent can be achieved by launching into equatorial orbit. Some U.S. firms have expressed interest in using Alcântara, but no plans have yet been made final.

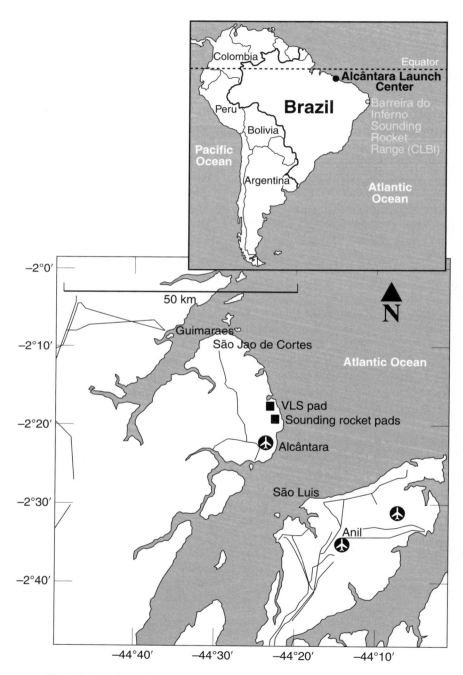

Fig. 2.2. Location of Alcântara Launch Center. (Site map © Mark Wade)

In November 2004, Brazil and Russia signed an agreement to cooperate on a number of space-related projects, including the joint utilization of Alcântara. The agreement did not specify whether the mutual use of Alcântara would also include launches of Russian Soyuz launch vehicles from the site.

Technical Facilities

Preparation Facility and Launchpad

These buildings include the preparation facility for vehicles and payloads, launchpad, and advanced control (blockhouse).

Operating Systems

The operating systems available in Alcântara include a radar system, telemetry system, control center, and meteorology facilities. Alcântara operates a 1 kW command and destruct system.

Radar System

The radar system consists of Adour and Atlas-Thomsom radar types. The Adour has a 3 m diameter antenna, frequency range from 5450 to 5900 MHz, peak power of 250 kW, and average power of 250 W. The Atlas has a 4 m diameter antenna, frequency range from 5450 to 5825 MHz, peak power of 1 MW and average power of 1 kW.

Telemetry System

The telemetry system operates in S and P bands. The S band has a 10 m diameter antenna, frequency range from 2200 to 2290 MHz, four receivers (two double chain carriers), decommutation PCM, PAM, and FM. The P band has a small antenna, frequency range from 300 MHz, and two receivers (one double chain carrier).

Control Center

The control center includes all facilities to control the operations during the countdown.

Satellite Control Station

This facility consists of a ground station, a satellite preparation complex, a satellite filling and assembly building, a chemical laboratory, and storage for pyrotechnics and propellants.

Payload Accommodations

Payloads can reach Alcântara by air, land, or sea. Special buildings can be used for storage and preparation of vehicles up to 50 tons, as well as facilities for payloads weighing about 2000 kg. Standard electrical power is available in 220/380 V, 60 Hz. An emergency power system can supply power on demand to specific sectors. A data-processing network assures accuracy and expedience to operations. A time generation and distribution system guarantees general time synchronization for the center. The site can accommodate 188 persons during a launch campaign.

A management support system is available, including interphone networks, specialized telephones, and operating signaling. The center provides facilities for both national and international telecommunications, such as long-distance calls, fax, and telex.

Acronyms

INPE   Instituto Nacional de Pesquisas Espaciais (National Institute for Space Research)
ISS   International Space Station
MECB   Missão Espacial Completa Brasileira (Complete Brazilian Space Mission)
VLS   Veículo Lançador de Satelites

Points of Contact

General coordination:
Departamento de Pesquisas y Desenvolvimento
Esplanada dos Ministérios–Bloco M
Edifício Anexo do MAer, 3°Andar
70.045-900–Brasília–DF Brasil
Fax: +55 61 224 6123

Technical and operational information:
Centro de Lançamento de Alcântara
CEP: 65.250-000–Alcântara–MA Brasil
Fax: +55 98 211 1069

References

AEB (Brazilian Space Agency), home page, 2003, <http://www.aeb.gov.br/centrolanc.htm> (17 January 2005).

"Alcântara Launch Center," <http://www.globalsecurity.org/space/world/brazil/Alcântara.htm> (17 January 2005).

F. D. Braun, "Houston, We Have a Brazilian," *Brazzil Magazine*, September 2003, <http://www.brazzil.com/p26mar98.htm> (17 January 2005).

"Brazil's Difficult Road to Space," *Space Today Online*, 2003, <http://www.spacetoday.org/Rockets/Brazil/BrazilRockets.html> (17 January 2005).

INPE (National Institute for Space Research) home page, 2001, <http://www.inpe.br/english/index.htm> (17 January 2005).

MECB (Brazilian Complete Space Mission) home page, <http://www.inpe.br/programas/mecb/ingl/default.htm> (17 January 2005). [Site contains photo gallery.]

"Spaceports Around the World: Brazil," *Space Today Online*, 2004, <http://www.spacetoday.org/Rockets/Spaceports/LaunchSites.html#Alcantara> (17 January 2005).

M. Wade, ed., "Alcantara," *Encyclopedia Astronautica*, 2004, <http://www.astronautix.com/sites/alcntara.htm> (17 January 2005).

Barreira do Inferno Sounding Rocket Range
• suborbital • 5.92 deg south • 35.15 deg west

Site Description

The Barreira do Inferno Sounding Rocket Range (Centro de Lançamento da Barreira do Inferno, or CLBI) was the first operational launch site in South America. It offers testing, launching, tracking, and data-acquisition operations to both Brazilian and foreign sounding rocket programs. The early series of Brazilian Sonda sounding rockets was testlaunched from the site.

CLBI is located in northeastern Brazil, about 17 km south of Natal, the capital city of the state of Rio Grande do Norte. The range has an area of approximately 18 km, bordered on the east by 6 km of the Atlantic coast. (Barreira do Inferno—literally, "Barrier of Hell"—takes its name from the colors of adjacent seaside cliffs.) This location was selected after the Ministry of Aeronautics drew up a list of site-selection criteria that included proximity to the equator, a large and undeveloped surrounding impact area, favorable prevailing winds, and preexisting logistical support (which was available in nearby Natal). CLBI's location affords excellent safety conditions and allows users to select from a broad array of launching azimuths, ranging from 14 to 145 deg.

Site construction began in 1964 with funding from the National Research Council and the Ministry of Aeronautics. Initial training support was provided by NASA's Goddard Space Flight Center. The range became operational in 1965 with the launchings of *Sonda-1* and Nike-Apache sounding rockets, and since then CLBI has seen hundreds of additional launchings. Sounding rockets launched here include *Sonda-2, -3*, and *-4*; Black Brant; Castor Lance; Super Loki; and *Skylark-12*.

A number of meteorological campaigns have been carried out at CLBI, including Project Ozone, a study of the ozone layer that began in 1978. Since 1979, as the result of an agreement with the European Space Agency (ESA), the range's facilities have been upgraded to enable CLBI to support Ariane rocket launchings, which take place at the Guiana Space Centre in neighboring French Guiana. The support consists of radar and telemetry data acquisition, real-time data processing, and transmission to the Guiana Space Centre and the Centre Spatiale de Toulouse in France. CLBI also offers support services for launchings that take place at Alcântara Launch Center, Brazil's orbital launch site. The organizational structure of CLBI comprises three divisions (Operations, Logistics, and Administrative) and two offices (Quality Assurance and Range Safety).

Technical Facilities

Launch Complex and Facilities

Launchpads

- Pad 1: A circular concrete area, 24 m in diameter, equipped with a universal rail launcher.
- Pad 2: A rectangular concrete area (20 × 6 m), equipped with a multirocket launcher.
- Pad 3: A circular concrete area, 20 m in diameter, equipped with a Nike-type launcher, which is also used to fit the Super Loki launcher adapter.
- Pad 4: A circular concrete area, 10 m in diameter, equipped with a Loki tubular launcher and a test rocket launcher.
- Pad 5: A rectangular concrete area (30 × 5 m), used to place users' mobile launchers.
- Pad 6: A rectangular concrete area (50 × 10 m), equipped with the *Sonda-4* assembly tower and launch platform.

Blockhouse

This is a circular structure, partly underground, built with reinforced concrete, with 12 protected windows that allow direct visual observation of the launchpads. It has approximately 36 sq m of space available for users' equipment.

Payload Assembly Building

This building has four rooms and a total available area of 180 sq m. Grounding system, communications, temperature control, and many other types of support equipment and systems are also available.

Vehicle Assembly Building

This is a 120 sq m building constructed in reinforced concrete. All types of ground support equipment for rocket checkout and assembly are available, such as hoists, dollies, forklifts, and air compressors.

Pyrotechnics Assembly Building

This is a 28 sq m building constructed and equipped in accordance with safety regulations.

Storage Buildings (2)

One (93 sq m) is for storage of rocket motors, the other (500 sq m) for nonhazardous materials.

Instrumentation

Radar System I

The Bearn is a C-band, scanning, tracking radar, with 4 m diameter antenna, frequency range of 5450–5825 MHz, and 1 kW peak power.

Radar System II

The Adour II is a C-band, scanning, tracking radar, with 3 m diameter antenna, frequency range of 5450–5825 MHz, and 250 kW peak power.

Telemetry System

The telemetry system is an S-band system with 10 m diameter antenna, frequency range of 2200–2300 MHz, six receivers, PCM decommutation, PAM, and FM/FM.

Data System

Tracking management and control are accomplished by two SOLAR 16/85 computer systems interfacing the radars and telemetry stations. In addition to controlling trajectory data acquisition and tracking, the systems furnish recorded trajectory data. The telemetry ground station has dedicated equipment that performs its internal data processing.

Meteorology Station

The meteorology station is designed for altitude and surface observations.

Time Generation System

This system ensures general time synchronization and distribution of coded time for all the operational stations.

Communications System

This system provides intrastation and external voice, signaling, and data communications. Intrastation communications are assured by a cable network, and external communications are provided by the Brazilian Telecommunications Company.

Command/Destruct System

CLBI operates a 200 W output power command/destruct system.

Management and Control

Functions performed at the control center are supported by a signaling system that displays the status of the stations and enables the go/no-go circuits.

Power Supply System

A redundant power supply is available for all the stations by commercial and locally generated power. The CLBI power plant consists of two 330 kV A generators (primary and backup).

User Support Services

In addition to operational services, CLBI offers two types of user support services, standard and optional.

Standard services:
- preoperational support (authorization, official documentation, reservations, etc.)
- customs clearance
- local transportation of material and personnel
- material storage
- workshop support
- health assistance

Natal's international airport is suitable for operation of all types of transport aircraft, providing parking, refueling, and flight operations services.

Optional services:
- lodging
- meals
- car rental
- material acquisition
- equipment repair
- aircraft support (parking, refueling, and maintenance)
- air cargo

Payload recovery service is also available.

Acronyms

CLBI	Centro de Lançamento da Barreira do Inferno (Barreira do Inferno Sounding Rocket Range)
ESA	European Space Agency
NASA	National Aeronautics and Space Agency (U.S.A.)

Point of Contact

Centro de Lançamento da Barreira do Inferno
RN 063
kM 11 C.P. 640
59022-970 Natal
RN Brasil
Voice: +55 84 211 0945
Fax: +55 84 211 4226

References

AEB (Brazilian Space Agency), home page, 2003, <http://www.aeb.gov.br/centrolanc.htm> (17 January 2005).

"Barreira do Inferno Launch Center," globalsecurity.org, 2003, <http://www.globalsecurity.org/space/world/brazil/inferno.htm> (17 January 2005).

Centro de Lançamento da Barreira do Inferno, <http://pessoais.digi.com.br/~clbi> (inactive).

M. Wade, ed., "Natal," *Encyclopedia Astronautica*, 2004, <http://www.astronautix.com/sites/natal.htm> (17 January 2005).

3 Canada

Churchill Rocket & Research Range
• suborbital • 58.8 deg north • 94.1 deg west

History

The Churchill Rocket & Research Range has been the focal point of the Canadian space program since the early 1950s. The range is located in northeastern Manitoba, close to the town of Churchill on Hudson Bay. Launch activity began in 1954 with the launch of a series of sounding rockets by the Canadian army at nearby Fort Churchill. Originally known as the Churchill Launch Range, the range became one of the world's earliest launch sites for scientific rockets following its dedication in 1957 and has since been the site of more than 3500 small-rocket launches.

Construction of the Churchill Range commenced in 1956, and the formal opening took place in 1957. Completion of the construction was timed to coincide with scientific experiments conducted worldwide throughout International Geophysical Year (IGY), July 1957–December 1958. Canadian scientists studied the near-Earth space environment during IGY by launching some 200 experiments onboard Nike-Cajun and Aerobee sounding rockets. The Churchill Range was closed at the end of IGY but was reopened in 1959 at the request of the U.S. Army.

A fire in 1960 destroyed many of the range's facilities. Following the transfer of Churchill to the U.S. Air Force in 1962, the launch site was rebuilt. The Canadian space program achieved a major milestone that year, when Canada became only the third country in the world to design and build its own satellite, the *Alouette I*. The U.S. Air Force stopped using Churchill in 1970, and the National Research Council of Canada assumed responsibility for the range's operations. Today Canada's only rocket range is operated by the Canadian Space Agency, which was established in 1989.

As a result of Canadian government cutbacks in late 1984, Churchill was closed as a launch facility. In September 1993 SpacePort Canada, a private company, was issued a lease for the operation of the range as a commercial launch complex. However, despite signing an agreement with Russia in 1996 to launch Start rockets into orbit, the company ceased operations in 1998 after several failed attempts to attract additional launch contracts and major investors.

The range is now owned by the province of Manitoba and is formally known as the Churchill Rocket & Research Centre. The site has been inactive since April 1998, when the last sounding rocket was launched. Some maintenance operations are currently contracted to the adjacent Churchill Northern Studies Centre to ensure that the range remains operational. The Churchill Rocket & Research Centre Committee is exploring possible commercial uses of the range.

Launch Site Description

The Churchill Launch Range is available for most sounding rockets and other small rockets up to 4300 kg gross weight. Situated far from urban and population centers, it offers an excellent location for launch operations. The current

Fig. 3.1. An Aerobee-Hi research rocket being fired at the IGY rocket launching site in Fort Churchill, Manitoba, Canada. (Courtesy National Academy of Sciences [U.S.])

environmental approval for operation of the launch range allows use of the most common propellants.

The Canadian Launch Safety Office oversees the country's established launch authorization process. This office provides a single launch-approval interface for potential users.

Fig. 3.2. Location of the Churchill Rocket & Research Range in Manitoba, Canada.

The sounding rocket launch facilities and equipment have supported a very high rate of successful launches for a number of years. The facilities at the range provide protection against the environment even during the coldest months. Passageways connect the various facilities to enable use in extreme weather.

Technical Facilities

Two operational launch buildings are capable of launching solid-motor sounding rockets of various sizes: the universal launch building (Pad 1) and the Aerobee launch tower (Pad 3). Several nonoperational launch buildings are used for storage. All launch complexes share a blockhouse that serves as the launch control center.

Pad 1

This concrete building was designed to launch sounding rockets from a number of manufacturers. Sounding rockets launched from this building include Nike-Tomahawk; Black Brant (*VB, VC, VIII, IX*, and *X*); Taurus-Orion; and Nike-Orion.

The launcher, a *D-8* manufactured by the AeroLab Development Co., has been well maintained and is still operational. It handles rockets up to 18 m in length and approximately 4500 kg in weight. The launcher's heat shield protects the rocket stages and payload from inclement weather.

Launcher azimuth, elevation, heat shield, and roof door operation can be controlled locally from the launch bay or remotely from the blockhouse.

The universal launch building is heated to facilitate work during cold weather. Its dimensions are 15.8 m (width) by 24.9 m (length) by 7.6 m (height).

Pad 3

The Pad 3 building was designed for launch of liquid-fueled Aerobee sounding rockets. It has been retrofitted to launch 43 cm diameter, three-fin vehicles, such as the Black Brant series of sounding rockets. The building has H-beam, steel-frame construction, and its walls are of sandwich construction with insulation between aluminum sheets. It sits on a concrete slab, and its dimensions are 14.6 m (length) by 14.6 m (width) by 15.8 m (height). An access door is 2.4 m wide by 3.0 m high.

The Pad 3 launch tower is located in a pit area in the center of the Pad 3 launch building and extends through its roof. The tower, 34.2 m high, is constructed of structural steel. It has five steel working platforms located inside the building supported on the launch tower. The Pad 3 tower is fitted with three launch rails spaced at 120 deg; these are capable of launching Black Brant VB, VIIIB, IXB, and XB sounding rockets. The tower loading rail is approximately 9.1 m long, but vehicles longer than that can be loaded in stages by raising the payload first with the hoist located on the fourth level. The tower rail configuration limits the maximum diameter of the rocket to 0.48 m.

Hazardous Assembly Building

Sounding rockets are assembled for launch in the hazardous assembly building. The operations performed in this facility may include final assembly of rocket stages, payload mating to the rocket, and igniter installation. The building, constructed of reinforced concrete, has a pyrotechnic storage vault, loading vestibule, assembly room, and two small offices. A 10-ton crane is located inside the assembly room.

Payload Preparation Building

Linked to Pad 3 via tunnel, this building is conditioned to support a shirtsleeve work environment. Workbenches, electrical power, and lighting are provided to support processing. A floor crane with 1.3 metric ton capacity is also available.

Blockhouse

This facility houses the test conductor, range safety officer, and payload control and mission controller control-panel positions. It contains the necessary patch-boards and terminal boxes for the instrumentation cabling that connects to the launchers and the telemetry room. Two user groups, as well as the equipment they require for normal launch operations, can be accommodated at the same time in the blockhouse. Constructed of reinforced concrete, the blockhouse has a concrete slab foundation. Its roof is composed of sand filled between two reinforced concrete slabs. It is heated year-round by an electrical three-phase heating system. The blockhouse's interior dimensions are 9.7 m wide by 16.3 m long by 6.5 m high from the top of the finished concrete slab.

Point of Contact

Churchill Rocket & Research Centre Committee
Room 648-155 Carlton Street
Winnipeg, Manitoba
Canada
R3C 3H8
Voice: (204) 945-8193
Fax: (204) 945-8229

References

"Canada's Churchill Spaceport," *Space Today Online*, 2003, <http://www.spacetoday.org/Rockets/Spaceports/Canada.html> (17 January 2005).

Churchill Launch Range, <www.churchill-launch.com> (inactive).

Friends of CRC Association, 2004, <http://friendsofcrc.ca/> (17 January 2005).

M. Wade, ed., "Fort Churchill," *Encyclopedia Astronautica*, 2003, <http://www.astronautix.com/sites/forchill.htm> (17 January 2005).

4 China

Overview

On 15 October 2003, the People's Republic of China joined the ranks of the world's space superpowers when the first "taikonaut" (Chinese for astronaut or cosmonaut), Lt. Col. Yang Liwei, and his *Shenzhou 5* spacecraft were launched into orbit. Yang returned to Earth after a 14-orbit, 21-hour successful mission that made China only the third nation, following Russia and the United States, to achieve human spaceflight. His flight capped an 11-year effort by China's space program to complete a crewed orbital spaceflight.

China's interest in space exploration dates back to the earliest years of the space age, but its space program has been marked by several starts and halts. The most recent full-scale effort was initiated in 1992, after China had embraced Western-style economic reforms. Russia provided both technical assistance and training for Chinese astronauts beginning in 1995. The current Chinese crewed space effort was formally inaugurated in 1999.

With the assistance of the Soviet Union, China had begun a program during the late 1950s to develop its own intermediate range ballistic missile (IRBM). After the Soviet Union halted assistance in 1960, the Chinese continued work on their indigenous missile program. Development work began in 1965 on the first Chinese intercontinental ballistic missile (ICBM), the *DF-5*, which later became the basis for the family of Long March (Changzheng) launch vehicles.

On 24 April 1970, China became the fifth nation to launch a satellite (originally *"Mao 1"* but now usually referred to as *DFH-1*) into orbit, following the Soviet Union, the United States, France, and Japan. The *Long March 1 (CZ-1)* rocket was the launch vehicle for the satellite, which transmitted the song "The East is Red" back to Earth. Early successful satellite launches spurred the Chinese to devise plans for an ambitious crewed space program in the 1970s, but the chaos of the Cultural Revolution proved too disruptive for the program to be implemented.

The crewed space program was largely abandoned, and during the 1980s the Chinese space effort concentrated on the development of uncrewed spacecraft and began to use its range of Long March launch vehicles for commercial purposes. The China Great Wall Industry Corporation began handling Chinese commercial space activities in 1985, and the first launch of a foreign satellite using a Long March rocket took place in 1990. A number of American satellites were launched using Chinese launch vehicles during the 1990s, until the issue of export controls forced a halt to the use of Chinese facilities by U.S. firms in 1998.

In November 2004, China announced that its second crewed spaceflight, scheduled for launch in late 2005, would be a 5–6 day mission. Two astronauts will be carried into orbit onboard the *Shenzhou 6* spacecraft. Additional plans for possible space projects include landing an uncrewed craft on the moon by 2010 and returning soil samples from the lunar surface by 2020. China's success has stirred both India and Japan to accelerate their own space efforts.

China has three major launch sites: the Jiuquan Satellite Launch Center (JSLC), the Xichang Satellite Launch Center (XSLC), and the Taiyuan Satellite Launch Center (TSLC).

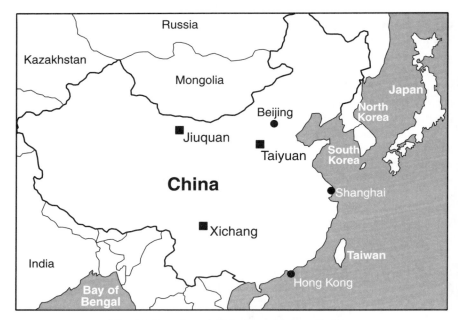

Fig. 4.1. Location of China's major launch sites.

Jiuquan Satellite Launch Center
• orbital • 41.3 deg north • 100.3 deg east

Launch Site Description

The Jiuquan Satellite Launch Center (JSLC) was established as the first Chinese rocket test facility, with construction beginning possibly as early as the late 1950s. JSLC is sometimes referred to in the West as Shuang Cheng-Tzu, the name of the nearest city. By 1964, its launch facilities were in place to support the first launch of a ballistic rocket. JSLC was the site where the first Chinese ICBMs were developed, and it was the launch site of the first Chinese satellite in 1970.

All Chinese spacecraft were launched from Jiuquan until 1983. JSLC was originally intended to provide low and medium Earth orbit launch capability for recoverable and scientific satellites. It is used for launching Long March (*CZ-2C, CZ-2D, CZ-2E,* and *CZ-2F*) launch vehicles and also sounding rockets.

After Chinese commercial space activities began in 1985, JSLC was sometimes used for commercial launches. However, because of its geographical constraints, most commercial activities are currently focused at China's other two principal launch sites. JSLC was the launch site for China's 2003 crewed orbital space mission.

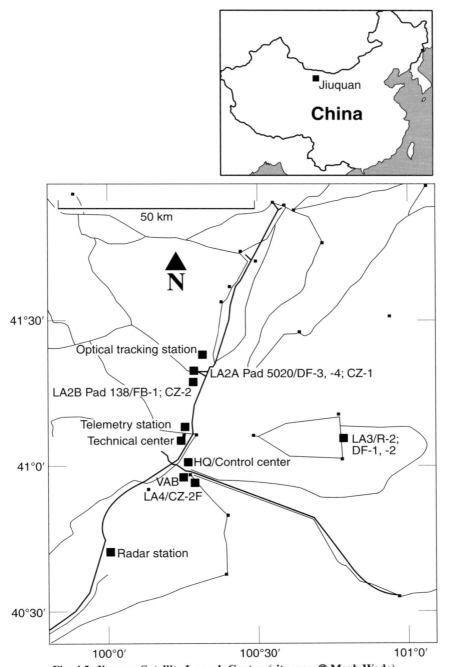

Fig. 4.2. Jiuquan Satellite Launch Center. (site map © Mark Wade)

JSLC is located near Jiuquan, at the southern edge of the Gobi Desert in Gansu Province in northwestern China. The rangehead is located about 50 km northeast of Shuang Cheng-Tzu, about 75 km from the Jiuquan airport. A rail line leads directly from Jiuquan to the launch site. JSLC's remote location was originally chosen because of the secretive nature of early test launch activity and the need for a place with low population density.

Launch corridor azimuths are constrained from 135 to 153 deg to avoid overflight of Russia and Mongolia. Typical missions are launched southeasterly into LEO (low Earth) orbits between 57 and 70 deg inclination. The orbital launch complexes are oriented toward the southeast, and the downrange instrumentation is also oriented in that direction. The desert terrain to the west allows the firing of surface-to-surface missiles within ranges of up to 1100 nautical miles inside Chinese territory.

Facilities

JSLC's facilities provide support to every phase of a satellite launch campaign. They include technical centers, launch complexes, launch control centers, a mission command and control center, propellant fueling system, communications systems, a gas supply system, a weather forecast system, and a logistics support system. The principal support area for operations is about 90 km south of Dong Feng (East Wind), the main base area.

A large instrumented area is dispersed along a 48 km stretch of the Etsin River, comprising a surface-to-surface missile (SSM) launch area, a surface-to-air missile (SAM) launch area, a large main support base containing nearly 200 buildings, a smaller support base servicing the SSM and SAM complexes, a large SSM and SAM assembly area, two storage areas, and several smaller housing and support areas.

JSLC had two launch complexes, launch areas 2 and 3, until 1999, when launch area 4 was opened. Launch area 2 consists of two launchpads (LA-2A and LA-2B) about 300 m apart with a shared service tower. A large rail-mobile gantry provides service to both launch positions. Each launch position comprises a launch stand, aperture, and single exhaust duct. Each position is attended by an umbilical tower with rotating service platforms.

Launch vehicles are assembled in a technical center about 30 km from Dong Feng. There are no provisions for environmental protection of the launch vehicle at either LA-2A or LA-2B; the mobile gantry provides this service. During the on-pad integration operations, crane service is provided by the mobile gantry for launch vehicle and payload lift operations. In November 2003, a new launch complex, LA-2S, became operational with the launch of an *LZ-2D* vehicle.

Launch area 4 (LA-4) has its own technical center and launch center, a mobile launchpad, a 75 m high umbilical tower, and a vertical processing building. LA-4 is used primarily to support China's crewed spaceflight program.

Xichang Satellite Launch Center
• orbital • 28.25 deg north • 102.04 deg east

Launch Site Description

XSLC was constructed to serve as the site for Chinese geosynchronous orbital launches, utilizing the Long March family of launch vehicles. The site was selected from 16 possible locations, because the Jiuquan launch site was not suitably located to serve as a base for geosynchronous launches. Construction was started in a canyon site at the foot of Mt. Liang Shan on the first launchpad, Launch Complex (LC) 1, in 1978 and was completed in time for the inaugural launch of the *Long March 3* rocket in January 1984. Following China's entry into the commercial launch services market, additional payload handling facilities were constructed to support foreign spacecraft handling operations. In addition, modifications were performed to the on-pad service structure to provide a clean-room environment for commercial payload mating operations.

After a series of successful domestic geosynchronous satellite launches, the first Chinese commercial launch of a satellite, the *Asiasat 1*, occurred on 7 April 1990. As the size and mass of commercial spacecraft increased, a second pad, LC 2, was designed for a follow-on *Long March 2E* vehicle. This pad, constructed in about 14 months, is located about 400 m from Launch Complex 1. LC 2 was operational for the inaugural flight of the Long March 2E in July 1990. In January 2000, LC 2 served as the launch site for *Zhongxing-22*, the first Chinese military communications satellite. Most commercial satellites, including communication, broadcasting, navigation, and scientific research satellites, have been launched from XSLC.

The most southerly of the Chinese launch sites, XSLC is located in the sub-tropical, mountainous Xichang region of Sichuan Province, in southwestern China, approximately 1800 m above sea level. Temperatures annually range between 7 and 26°C, with an average of 16°C. Winds in this region are generally low throughout the year. The wet season lasts from June to September; the dry season takes up the rest of the year.

The main office of the launch center is located in the city of Xichang, some 65 km from the launch site. The Xichang airport, capable of handling jumbo aircraft such as C-130s and Boeing 747s, is 50 km from the launch site, and a dedicated railway and highway also lead directly to the XSLC. The center comprises the two previously mentioned launch complexes along with associated support facilities.

XSLC supports all geostationary missions from its location in southern China. The nominal launching azimuth is 97 deg, with downrange safety constraints limiting launch azimuths to a range of 94 to 104 deg. The Xichang Center is under the control of China Satellite Launch and Tracking Control General (CLTC) in Beijing. The Command and Control Centre, 7 km southwest of the launch site, is

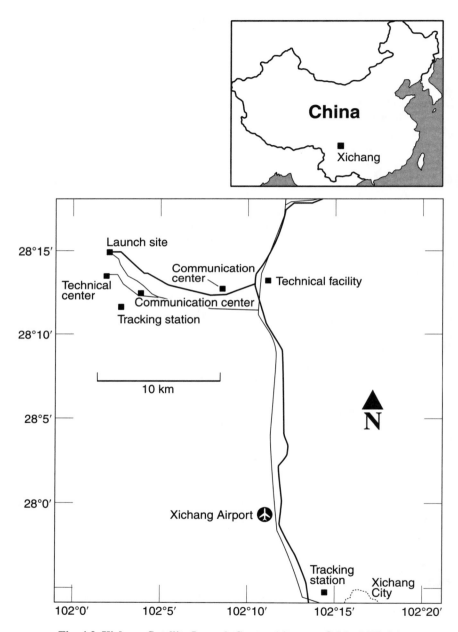

Fig. 4.3. Xichang Satellite Launch Center. (site map © Mark Wade)

the heart of control during system rehearsal and launch operations. Other facilities include communication systems to provide telephone and data communications, as well as meteorology support equipment. Two separate launchpads support flight operations: *CZ-3* is launched from Launch Complex 1, while *CZ-2E*, *CZ-3A*, *CZ-3C*, and *CZ-3E* are launched from Launch Complex 2. During 1993–1994, Xichang underwent extensive modernization and expansion, in part because of the requirements of the *CZ-3A/B/C* family and in part to meet commercial customer needs.

Facilities

The facilities at XSLC provide support to every phase of a satellite launch mission. The technical center is fully equipped for testing and integration of the payload and launch vehicle. The mission command and control center (MCCC), located 7 km southwest of the launchpad, is responsible for flight and safety control during overall system rehearsal and launch. The MCCC building is divided into two parts: the command and control hall, and the computer room. The command and control hall consists of the command area and the safety control area, and it contains operation rooms and offices. A viewing room on the second floor enables visitors to view launches on closed-circuit television screens.

Additional XSLC facilities include the propellant fueling systems, communications systems for launch command and for providing telephone and data communications to customers, and support equipment for meteorological monitoring and forecasting. Downrange telemetry, tracking, & control (TT&C) stations of XSLC are located in the cities of Xichang and Yibin in Sichuan Province, and in Guiyang City in Guizhou Province.

XSLC provides services for transportation, propellant storage, gas supply, RF relay link, and communications. For foreign spacecraft launches, satellite communications (including telephone, telex, fax, and data transmission) are used to link XSLC to the customer's home country. Dedicated duplex communication links can be established via two paths between XSLC and the user's country. During the mission, only one path is used while the other serves as a backup link. Communication service includes

- Audio communications: Telephone and facsimile communications can be established by using the secretary's console and facsimile machine via the dedicated circuitry of the XSLC.
- Data communications: Transmission rate of international data communications is 2400 bps. The maximum rate is 4800 bps.
- Telex: Communications can be established by using the telegraph automatic switching system.

Technical Center

The technical center, consisting of the launch vehicle preparation facilities and the payload preparation facilities, is located 2.2 km from the launch complex.

The launch vehicle and the payload are processed, tested, checked, assembled, and stored here. The launch vehicle preparation facilities consist of the transit building (BL1) and the testing building (BL2). A railway branch leads directly into the launch vehicle preparation building, with a turnaround line of 260 m, and a loading dock and freight house are also available. The payload preparation facilities include the nonhazardous operation building (BS2), the hazardous operation and fueling building (BS3), the solid rocket motor checkout and preparation building (BM), and the solid rocket motor X-ray building (BMX). Each houses tracking and measurement equipment for the powered phase of a launch vehicle flight.

Taiyuan Satellite Launch Center
• orbital • 39.17 deg north • 111.83 deg east

TSLC, also referred to as the Wuzhai Missile and Space Test Center by the U.S. Space Command, is located in Shanxi Province in northeastern China, about 280 km from the city of Taiyuan. The location was selected to provide China with domestic sun-synchronous and polar satellite launch capabilities. TSLC conducted its first missile test in 1979 and was initially used for testing rockets and missiles that were too large to launch from Jiuquan. Tests of the *DF-5* ICBM took place at TSLC between January 1979 and December 1981. The *DF-31* missile has also been tested here, and TSLC continues to be a major missile-testing facility for the Chinese military. TSLC currently contains a single space or orbital launchpad.

TSLC became an operational satellite launch facility in September 1988 by supporting the inaugural launch of the *Long March 4* for remote-sensing, meteorological, and reconnaissance missions. The second launch from TSLC, with an onboard weather satellite, occurred in September 1990. Satellite launches have taken place on a regular basis since then, the most recent a *ZY-2C* launched by a *CZ-4B*. This launch was a joint Chinese-European mission to study Earth's magnetic field. The launch corridor azimuths are constrained to a range from 90 to 190 deg. TSLC is currently offered as a commercial launch facility for satellites utilizing the *Long March 4* launcher. During the late 1990s, several U.S. Iridium satellites were launched from TSLC onboard *Long March 2* rockets.

The main office of the launch center is located in Taiyuan City. TSLC has two feeder railways that connect with the Ningwu-Kelan railway, and TSLC has its own railway station. Highways also lead to all TSLC sites. Taiyuan Airport, 300 km from the launch site, has runway capacity for jumbo aircraft.

Taiyuan City is bordered by the Taihang Mountains in the east and the Luliang Mountains in the west, and the Fenhe River runs through the city from north to south. TSLC is located in a temperate climate zone, some 1400–1900 m above sea level, and has an annual average temperature of 9.5°C.

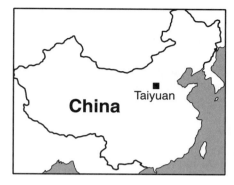

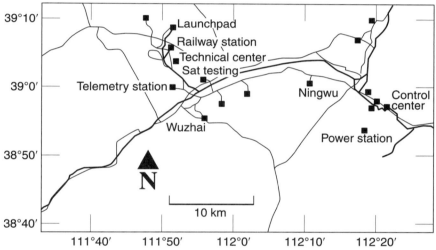

Fig. 4.4. Taiyuan Satellite Launch Center. (site map © Mark Wade)

Launch Site Description

The TSLC launch facility is patterned after the Xichang Satellite Launch Center site. It functions primarily to launch meteorological, reconnaissance, and remote-sensing satellites using Long March launch vehicles. The principal facilities of TSLC include the technical center; the mission command and control center; a TT&C system; and associated technical and logistic support systems, including the center's own power station.

The single launchpad comprises a launch mount, exhaust aperture, and a single below-grade exhaust duct. The launch position is serviced by a fixed service tower with rotatable service platforms. A large service crane is mounted atop the tower that is utilized for lift operations during the on-pad integration process. Lightning protection is provided by lattice-type conductor towers around the perimeter of the site. The *Long March 2C/SD* and *4B* are the active launchers currently associated with TSLC. In September 2002 and September 2003, TSLC was the site of two *KT-1* vehicle launches, but both were failures.

Details of the TSLC technical center for launch vehicle and payload processing have not been publicized to date, but launch vehicle processing at TSLC is probably similar to that conducted at other Chinese launch sites. Payload processing is available for commercial launch operations.

Acronyms

CLTC	China Satellite Launch and Tracking Control General
ICBM	intercontinental ballistic missile
IRBM	intermediate range ballistic missile
JSLC	Jiuquan Satellite Launch Center
LC	launch complex
LEO	low Earth orbit
MCCC	mission command and control center
SAM	surface-to-air missile
SSM	surface-to-surface missile
TSLC	Taiyuan Satellite Launch Center
TT&C	telemetry, tracking, and control
XSLC	Xichang Satellite Launch Center

Points of Contact

China Great Wall Industry Corporation
30 Haidian Nanlu,
Beijing 100080,
P.R.China
Voice: 86-10-68748888/68748810
Fax: 86-10-68748876/68748865
E-mail: cgwic@cgwic.com
Web: www.cgwic.com

G W Aerospace Inc.
21515 Hawthorne Blvd Suite 1065
Torrance, CA 90503-6518
U.S.A.
Voice: (310) 540-7706

References

Chen Lan, *Go Taikonauts!: An Unofficial Chinese Space Web Site*, last modification 27 September 2004, <http://www.geocities.com/CapeCanaveral/Launchpad/1921/> (25 March 2005).

China National Space Administration, <http://www.cnsa.gov.cn/main_e.asp> (25 March 2005).

"China's Space Program," *SpaceRef.com*, 2005, <http://www.spaceref.com/focuson/china/> (25 March 2005).

S. J. Isakowitz, J. P. Hopkins Jr., and J. B. Hopkins, *International Space Launch Reference Guide* (AIAA, Reston, VA, 1999).

J. Pike, "Chinese Space Activities," *Federation of American Scientists: FAS Space Policy Project—World Space Guide*, <http://www.fas.org/spp/guide/china/> (25 March 2005).

"Space Rocket Launch Sites Around the World," *Space Today Online*, 2004, <http://www.spacetoday.org/Rockets/Spaceports/LaunchSites.html> (25 March 2005).

M. Wade, ed., "China," *Encyclopedia Astronautica*, last modification 28 March 2005, <http://www.astronautix.com/articles/china.htm> (30 March 2005).

M. Wade, ed., "Jiuquan," *Encyclopedia Astronautica*, last modification 30 March 2005, <http://www.astronautix.com/sites/jiuquan.htm> (30 March 2005).

M. Wade, ed., "Taiyuan," *Encyclopedia Astronautica*, last modification 30 March 2005, <http://www.astronautix.com/sites/taiyuan.htm> (30 March 2005).

M. Wade, ed., "Xichang," *Encyclopedia Astronautica*, last modification 30 March 2005, <http://www.astronautix.com/sites/xichang.htm> (30 March 2005).

5 France and French Guiana

Guiana Space Center
• orbital • 5.23 deg north • 52.75 deg west

Overview

The Guiana Space Center (Centre Spatial Guyanais, CSG) is governed under an agreement between the French space agency, the National Center for Space Studies (Centre National d'Études Spatial, CNES), and the European Space Agency (ESA); its day-to-day operations are managed by CNES on behalf of ESA. It is situated on the Atlantic coast of French Guiana, an overseas department of France in the northeastern part of South America. Its launch facilities are located approximately 65 km from Cayenne, the country's capital and largest city, on a strip of land between the towns of Kourou and Sinnamary, along the coastline. CSG is sometimes referred to as Kourou, because that town is the nearest one.

France's launch site was not always located in French Guiana. The French space program has a long history, with rocket research and development dating back to the years immediately following World War II. The French had developed a site at Hammaguir, Algeria, for the testing of rockets beginning in 1952. (Algeria was then still a French possession.)

In keeping with the French desire for independence from both the Soviet bloc and the American-led West, President Charles de Gaulle advocated an autonomous French space program—one that did not rely on the two space superpowers, the United States and the Soviet Union, for technical support. (France did, however, cooperate with both countries on projects beginning in the early 1960s.)

De Gaulle was one of the major backers of the creation of CNES as part of his program to assert French sovereignty by making France a major world power. CNES was established in 1961 to develop autonomous access to space for French launch vehicles. Today, CNES programs are focused on space technologies for commercial and environmental applications.

By the early 1960s the frequency of launchings at Hammaguir had greatly increased. On October 18 1963, the French successfully launched and recovered the first cat sent into space. "Felix" was launched from Hammaguir on a Véronique AGI sounding rocket.

Two years later, France became only the third nation to orbit its own satellite using its own launch vehicle, the Diamant. The satellite was dubbed *"Asterix 1"* after the popular French comic strip character. Subsequently, in 1982, the first French orbital mission flown by a cosmonaut would take place, when Jean-Loup Crétien would fly onboard the Soviet *Salyut 7* space station for eight days.

When Algeria won independence in 1962, CNES lost its launch site in that country. The agency selected the French Guiana site as the French national spaceport in April 1964, after investigation of 14 other possible sites. Criteria used in the final selection included proximity to the equator (an ideal location for placing satellites into geostationary orbit), lack of hurricanes and earthquakes in the region, low population density, ability to launch over the Atlantic Ocean, and availability of a number of launch azimuth possibilities ranging from east to north.

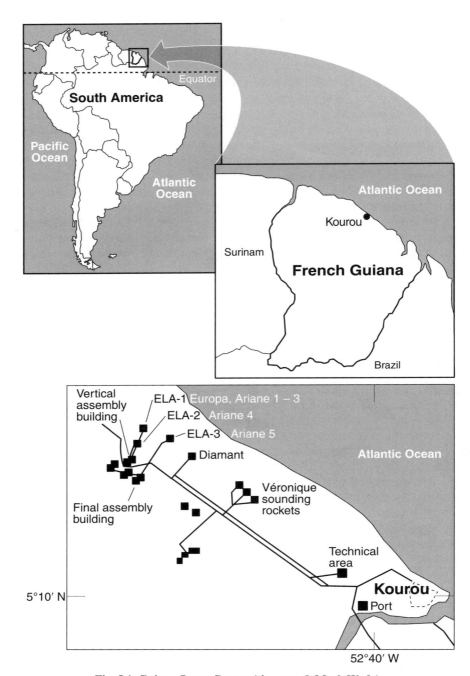

Fig. 5.1. Guiana Space Center. (site map © Mark Wade)

CSG became operational in April 1968 with the launch of a Véronique sounding rocket. The first orbital launch of a satellite from the spaceport, using a Diamant-B rocket, took place on 10 March 1970. By 1975, the year of the final Diamant launch, 184 sounding rockets had been launched from CSG, and nine orbital launches had taken place. Beginning that year, a major renovation of the Kourou facilities was undertaken to accommodate the needs of the proposed Ariane launch vehicle, which had been approved for development by the European Space Conference in 1973. This decision resulted from the European need for an autonomous launch vehicle, and the drive for independent access to space was formalized that year by the creation of the ESA.

CNES submitted a proposal to ESA to develop a new booster that would replace the failed *Europa II* launcher, the first European orbital launch vehicle, launched in November 1971. ESA undertook responsibility for management and financing of the new vehicle, but the technical development of the launcher was assigned to CNES. French membership in ESA and heavy French participation in the development of the Ariane rocket led to the adoption of CSG as the official spaceport of ESA after the French offered to share its facilities. As a result, CSG is known today as "Europe's spaceport."

Two-thirds of CSG's annual operating budget is contributed by ESA, which currently has 17 member states. General coordination of CSG is shared by CNES and ESA, while the Arianespace company serves as the launch operator and EADS Space Transportation is the prime contractor for the Ariane 5 launcher. These four companies are the principal participants in the overall operations of CSG. CNES is responsible for construction of CSG facilities, coordination of launch operations, and spacecraft processing services. ESA oversees the direction of the Ariane launcher program in addition to making contributions to the CSG budget to fund development of Ariane-related installations.

The first Ariane launch took place in December 1979 from the Ensemble de Lancement Ariane 1 (ELA-1) pad. Construction on the ELA-2 launchpad was completed in 1986 for launches of the *Ariane 4*. The final major CSG launchpad, ELA-3, was completed in 1996 for *Ariane 5* launchings. Today the operations of CSG are almost entirely focused on the Ariane launch vehicle, although investigations continue into the possibility of using CSG as a base for other launch vehicles, including the Russian Soyuz and the Ukrainian *Tsyklon-4*. In addition, CSG is constructing facilities for the first launch of ESA's Vega small launcher, which is currently scheduled for 2006. CSG has been extremely successful in recent years, capturing about 50 percent of the market for commercial satellites launched into geostationary orbit.

Launch Site Description

CSG covers a surface area of approximately 750 sq km. Its eastern boundary is marked by a 50 km coastline along the Atlantic. It is bordered on two sides by tropical forest and on the south by the town of Kourou, which has the closest port facilities. The first four launchpads constructed at CSG, built for sounding rockets, were completed in 1968. Additional pads for orbital vehicles were built for the Diamant-B rocket in 1969 and the *Europa II*, finished in 1971.

Pads constructed for launching the Ariane series of orbital vehicles are ELA-1, ELA-2, and ELA-3. ELA-1 was completed in 1979 by modifying the pad formerly used for the single launch of the *Europa II*, which occurred in 1971. The first launch of the *Ariane 1* from ELA-1 took place on 24 December 1979. ELA-1 was also used for the launches of the *Ariane 2* and *Ariane 3*, which were essentially modified versions of the *Ariane 1*.

The ELA-2 pad was completed in 1986 to support launches of the *Ariane 4*, which was first launched in June 1988. Construction began on ELA-3 that same year to support operations of the heavy-lift *Ariane 5*. ELA-3 was finished in 1996, and the first launch of an *Ariane 5* occurred on June 4 of that year, while the launch of the first *Ariane 5* carrying a commercial payload took place in December 1999.

ELA-3

With the completion of *Ariane-4* launches in 2003, the ELA-3 is currently the only active pad used for orbital launches, although ELA-1 is currently being modified for launches of the Italian-produced Vega rocket, scheduled to begin in 2006. Arianespace is responsible for the upkeep of ELA-3, but all the launchpads are the property of the ESA. ELA-3 has the capacity to reduce the time between launches to one month. The 3 by 5 km launch complex includes all facilities required for carrying out all necessary operations prior to the actual launch during the 22-day launch campaign. ELA-3 consists of two primary areas, the preparation zone and the Ariane launch zone (ZLA).

Preparation Zone

The preparation zone's principal buildings for launch vehicle integration are
* booster integration building (BIP)
* launcher integration building (BIL)
* final assembly building (BAF)

These buildings are widely separated because of safety concerns.

The *Ariane 5*'s two solid boosters are assembled and prepared in the 55 m high BIP. The forward, center, and aft segments are mated in the BIP's two integration rooms and fully inspected before being moved to the BIL.

In the 58 m high BIL, the cryogenic central core stage of the *Ariane 5* is positioned vertically over the mobile launch table (MLT). The MLT is a 900-ton rail-mobile structure that utilizes dual rail tracks to transport the launcher and launch platform. An umbilical tower attached to it eliminates the need for an on-site launch-pad tower. The two solid boosters are vertically docked to the rocket's main stage. The cryogenic upper stage and the vehicle equipment bay are mounted atop the central core to complete assembly of the *Ariane 5*'s lower composite. The launch vehicle's electrical, fluid, and mechanical systems are inspected before leaving the BIL.

Fig. 5.2. *Ariane 5* **at Launch Zone no. 3 at CSG. (Reprinted with permission of CNES.)**

After about 16 days in the BIL, the cryogenic stage with strap-on boosters is transported to the BAF, where it is mated with its payload. The mating operation usually requires two weeks. In the BAF, the upper composite of the launcher, which comprises the fairing and either one or two satellites, is mounted onto the lower composite. The rocket's upper stage and altitude control systems are then filled with propellants. When all final checks are completed, one day prior to lift-off, the fully integrated launch vehicle is transferred to the launch zone using the MLT. A tractor pulls the *Ariane 5* down the rail tracks at about 3.5 km/h.

Ariane Launch Zone

The ZLA contains the launchpad, which is connected to the preparation zone by MLT rail tracks. The Ariane 5 makes the 5 km journey from the BAF to the launchpad in about 90 min using the MLT. ELA-3 was constructed using the "clean pad" concept—the launch vehicle was assembled, tested, and mated with its spacecraft payload away from the pad to allow for greater flexibility in launch operations. With this procedure, the amount of support infrastructure is reduced, because only propellant loading and the actual launch occur at the launchpad.

The pad is surrounded by four 80 m high lightning-conductor towers for pro-tection during storms and a 90 m high water tower, which pours water onto the launch table during liftoff to help cool the table and the adjacent trenches. A nearby building was constructed to protect the electrical and fluid connections. The area near the launchpad also includes exhaust ducts and deflectors. During the launch countdown, the cryogenic tanks of the rocket's main stage are filled, the launcher's subsystems are inspected and activated, and the flight system is loaded.

Launch Operation Facilities

CSG has extensive support facilities for handling launch campaigns, and it pro-vides general logistics support during *Ariane 5* preparation and launch phases. A launch campaign for the *Ariane 5* typically lasts five weeks. Ground facilities at CSG used to support launches include optical systems, radar systems, a launcher telemetry acquisition system, a flight safety system, telecommunication systems, and a meteorological center.

The satellite preparation facilities complex handles the preparation of satellite payloads prior to their integration with the launcher. The Payload Preparation Complex consists of four major areas: S1, the Payload Processing Facility; S3, the Hazardous Processing Facility; S2-S4, Hazardous Processing Facility for solid motors and pyro devices; and S5, a combined set of Payload/Hazardous Process-ing Facilities. These areas provide similar capabilities, allowing simultaneous pro-cessing of multiple satellites. After all satellite systems have been checked out and the payload's client company has given final approval, the satellite is moved to the BAF for mating with the launch vehicle.

The Launch Control Centre (CDL3, "Centre de Lancement no. 3") monitors launcher status up to the launch.

Within the Technical Centre, the Mission Control Centre (located in the Jupi-ter building, about 12 km from the launchpad) receives and manages all prelaunch and launch information. The launch countdown takes place here, and the launch team monitors the *Ariane 5* and its payload until the satellite has been correctly placed in orbit, approximately 40 minutes after liftoff. The director of operations is responsible for overseeing the launch sequence and final countdown.

The 50 m high booster engine test stand is used to conduct inspection tests of launcher engines. The engines are ignited over a massive pit 60 m deep and 200 m long. CSG has two propellant plants, both operated by the Air Liquide company, for production of liquid hydrogen and liquid oxygen to fuel the *Ariane 5* engines.

The Guiana propellant plant (UPG) produces the solid propellant to power the Ariane booster. Once the three booster sections are received, the central and aft segments are loaded with 107 tons of solid propellant. The forward segment is loaded with propellant prior to being shipped to CSG. The UPG is a highly automated facility, with a production capacity of 16 boosters per year. Testing facilities for boosters are available in the UPG's solid-fuel acceleration test facility. The UPG is connected by rail to the BIP.

Acronyms

BAF	final assembly building
BIL	launcher integration building
BIP	booster integration building
CNES	Centre National d'Études Spatiales
CSG	Centre Spatial Guyanais
ELA	Ensemble de Lancement Ariane
ESA	European Space Agency
MLT	mobile launch table
UPG	Guiana propellant plant
ZLA	Ariane launch zone

Points of Contact

Arianespace Inc.
601 13th St. NW
Washington, D.C. 20005
U.S.A.
Voice: (202) 628-3936
Fax: (202) 628-3949

Arianespace
Boulevard de l'Europe
91006 Evry Courcouronnes
Cedex, France
Voice: +33-1-60-87-60-00
Fax: +33-1-60-87-63-04

References

"Ariane 5 User's Manual, Issue 4," *Arianespace*, 2004, <http://www.arianespace.com/site/documents/sub_main_ariane5_manual.html> (17 August 2005).

"Europe's Spaceport," *Arianespace*, 2005, <http://www.arianespace.com/site/spaceport/spaceport_index.html> (26 January 2005).

"Europe's Spaceport," *European Space Agency*, last update 6 May 2004, <http://www.esa.int/export/SPECIALS/Launchers_Europe_s_Spaceport/> (26 January 2005)

"Centre Spatial Guyanais," *Centre National D'Études Spatiales*, 2005, <http://www.cnes.fr/html/_1016_.php> (26 January 2005).

"CNES – The French Space Agency," *Centre National D'Études Spatiales*, 2005, <http://www.cnes.fr/html/_455_460_1153_.php> (26 January 2005).

S. J. Isakowitz, J. P. Hopkins Jr., and J. B. Hopkins, *International Reference Guide to Space Launch Systems* (AIAA, Reston, VA, 1999).

P. Redfield, *Space in the Tropics: From Convicts to Rockets in French Guiana* (University of California Press, Berkeley, 2000).

J.-J. Serra, "CSG (Guiana Space Center)," *Rockets in Europe*, <http://www.univ-perp.fr/fuseurop/kourou_e.htm> (24 March 2005).

M. Wade, ed., "Kourou," *Encyclopedia Astronautica*, 2004, <http://www.astronautix.com/sites/kourou.htm> (26 January 2005).

6 India

Satish Dhawan Space Centre
• orbital • 13.6 deg north • 80.25 deg east

Overview

India's national space program marked its 40th anniversary in 2003. In 1963, the first sounding rocket was launched from a range near Thumba, in the state of Kerala, only one year after the formation of the Indian National Committee for Space Research. After four decades, the Indian space effort is now widely regarded as especially successful in its applications of space technology to India's development and to improvement of the quality of life for the nation's estimated one billion people. As a result, India's space program has become a model for developing nations that aspire to create one but cannot compete with the major space powers. Despite a low (by American standards) annual budget of approximately U.S.$450 million, India has had notable successes with its launch campaigns; by the beginning of 2004, only six out of 39 satellite launches had failed. As a result, its space program has become a source of considerable national pride.

The Indian space effort is managed by the Indian Space Research Organization (ISRO), which was organized in 1969, initially as a division of the Department of Atomic Energy. In 1972, it was brought under the newly formed Department of Space. Just three years later, the first Indian satellite, Aryabhata, was launched via a Soviet rocket. India launched its *Rohini-1* satellite using its own launch vehicle, the Satellite Launch Vehicle (SLV), from the Sriharikota Launch Range on July 18, 1980. India's first cosmonaut, Rakesh Sharma, flew onboard the *Salyut 7* space station for eight days in 1984 as part of a joint Indo-Soviet mission. During the 1980s and 1990s, Indian scientists concentrated on the development of satellites for communications and remote sensing. Current operational Indian launch vehicles include the 294-metric-ton Polar Satellite Launch Vehicle (PSLV) and the 401-metric-ton Geosynchronous Launch Vehicle (GSLV). India hopes to eventually market the GSLV to commercial users.

The goals of India's space program will likely expand beyond primarily developmental applications following China's successful launch of an astronaut in October 2003. With India and China each striving to set an example of technological and scientific accomplishments for the developing world, the successful Chinese crewed mission seems to have spurred Indian space efforts. The first indication of this was a renewed call for sending an unmanned Indian probe to the moon, which was initially proposed in 2002. Some Indian scientists have also called for sending a satellite to Mars. A probe to either the moon or Mars would utilize the GSLV launch vehicle. Another group of scientists within India's space community advocates that the country's long-range space goals eventually include a crewed lunar landing. India had hoped to market some of its space-related technology to other nations to make its space program a profitable enterprise, but recent export attempts have been largely unsuccessful.

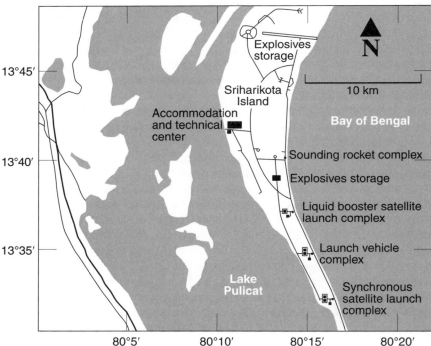

Fig. 6.1. Satish Dhawan Space Centre is on Sriharikota Island on the east coast of India, about 60 miles north of Madras.

Launch Site Description

India's primary launch complex is the Satish Dhawan Space Centre, formerly known as the Sriharikota Space Centre. The name was changed in 2002 following the death of Satish Dhawan, who had overseen the Indian space program since the death of its founder, Vikram Sarabhai, in 1971. The ISRO oversees the center's operations. The space center is located on Sriharikota Island on India's eastern coast, about 60 miles north of the city of Madras in the state of Andhra Pradesh. Its launch range, the Sriharikota Launch Range (SHAR), consists of facilities for the testing, preparation, integration, and launch of the PSLV and GSLV launch vehicles. SHAR covers an area of some 145 sq km. The launch range became operational in October 1971 with the launch of three *Rohini-125* sounding rockets. In addition to the SHAR, the ISRO also operates launch ranges at Thumba (near Trivandrum) and Balasore, for conducting sounding rocket flights for scientific studies.

The siting of the Satish Dhawan Space Centre makes north-south polar launches impossible because of nearby heavily populated areas. Polar-orbit missions must be launched in a southeastern direction, with launch azimuth limited to 140 deg. These missions also require an intensive 55 deg yaw maneuver following launch. Geosynchronous transfer orbit missions can be launched with a launch corridor at an azimuth of 102 deg.

SHAR's facilities include a mission control center and a large network of tracking radars. A 75 m, 3000 ton mobile service tower was constructed on the PSLV pad for launch vehicle integration. The tower provides a payload clean room at the 41 m level. A solid-propellant space booster plant processes large propellant grains for the launch vehicle and stage motor programs. The Static Test and Evaluation Complex is used for testing and qualifying solid motors used with the PSLV launch vehicle. The SHAR closed center contains real-time tracking systems, meteorological observation equipment, and computers linked to the ISRO's telemetry, tracking, and command network.

Range operation activities comprise mission analysis and range safety, programmed management, computer facilities, meteorology, and range instrumentation. Two major components at Sriharikota are the liquid propellant storage and service facilities and the vehicle assembly and static test facilities. A quality and reliability group and a programmed planning and evaluation group provide necessary support to the center. Sriharikota Common Facilities (SCF) looks after engineering maintenance, transport, fire services, medical care, and public health.

The two launch sites are used for launching the PSLV and the GSLV. The PSLV can place a 1200 kg satellite into orbit, while the GSLV is capable of lifting a 1900 kg payload to geosynchronous transfer orbit. The PSLV site, first commissioned in 1990, contains the mobile service tower. A new GSLV launchpad is also under construction, probably for the new GSLV Mark III launch vehicle now under development.

SHAR facilities consist of the following major elements:
• launch control center and mission control center

- satellite preparation facility
- mobile service tower, umbilical tower, and launch pedestal
- solid-motor preparation facility
- subsystem preparation facility
- hardware storage facility
- industrial service facility
- liquid/cryogenic propellant storage and transfer facility
- cryogenic stage preparation facility
- automatic checkout and control for propellant filling
- safety and firefighting systems
   Range instrumentation and support facilities include:
- TM data receiving station
- tracking systems, including precision 'C' band mono pulse radars and 'C' and 'S' band radars
- meteorological and technical photography
- telecommand support systems, including intercommunication, CCTV system, data links, range timing system, real-time systems, and specialist display system

Acronyms

GSLV	Geosynchronous Launch Vehicle
ISRO	Indian Space Research Organization
PSLV	Polar Satellite Launch Vehicle
SLV	Satellite Launch Vehicle
SCF	Sriharikota Common Facilities
SHAR	Sriharikota Launch Range

Points of Contact

Antrix Corporation Limited
Antrix Complex, New BEL Road, Bangalore, 560 094
India
Voice: +91-80-3415474
Fax: +91-80-3418981

Indian Space Research Organization
New BEL Road, Bangalore 560 094
India
Voice: +91-80-3415275 & 3415474
Fax: +91-80-3412253

Vikram Sarabhai Space Center
Trivandrum, 695022

Kerala, India
Voice: +91-471 562-444/562-555
Fax: +91-471-7979

Sriharikota Launch Range
Andhra Pradesh 524124
India
Voice: +91-2001-041-394
Fax: +91-2001-041-568-594

References

"India in Space," *Space Today Online*, 2004,
<http://www.spacetoday.org/India/India.html> (16 February 2005).

"India in Space – Indian Space Program," *Aerospaceguide*, 2005,
<http://members.lycos.co.uk/spaceprojects/world_space/india_in_space.html>
(16 February 2005)

"India Space Program Research," *Indian Child*, 2005,
<http://www.indianchild.com/india_space_research.htm> (16 February 2005).

Indian Space Research Organization, <http://www.isro.org/> (16 February 2005).

"India's Space Program Turns Forty," *Space.com*, 2005,
<http://www.space.com/missionlaunches/india_forty_031123.html> (16 February
2005)

S.J. Isakowitz, J.P. Hopkins Jr., and J.B. Hopkins, *International Reference Guide
to Space Launch Systems*, 4th ed. (AIAA, Reston, VA, 2004).

M. Wade, ed., "Sriharikota," *Encyclopedia Astronautica*, 2004,
<http://www.astronautix.com/sites/sriikota.htm> (16 February 2005).

7 Israel

Palmachim Launch Site
• orbital • 31.9 deg north • 34.7 deg east

Overview

Israeli launch activities have taken place entirely from what has come to be known as the Palmachim launch facility, named for nearby Palmachim Air Force Base. The Ofeq ("Horizon") 1–6 series of military reconnaissance satellites are the only spacecraft that have been launched from this site.

Israel's interest in space exploration dates back nearly to the beginning of the space age with the founding of the National Committee for Space Research (NCSR) in 1960. The space research conducted by NCSR scientists during the next two decades ultimately laid the groundwork for the establishment of the Israel Space Agency (ISA) in 1983. Organized by the Ministry of Science, ISA was founded to support and coordinate both academic and private-sector space research, oversee the integration of projects that involve multiple participants, and promote the importance of the space program to the Israeli public. Since its founding, ISA has been charged with coordination of the Israeli space effort, which focuses primarily on the launch of reconnaissance and communication satellites.

Israel formally entered the space age with the successful launch of its first satellite, *Ofeq-1*, into orbit on 19 September 1988. With this launch, Israel joined the exclusive ranks of the "space club," becoming only the eighth nation to launch its own satellite aboard its own launch vehicle, the solid-fuel, three-stage Shavit ("Comet") rocket.

A high point of the Israeli space program was the selection of Air Force pilot Ilan Ramon in 1995 to be the country's first astronaut. Ramon became a national hero when he flew onboard the *Columbia* space shuttle in January 2003—the mission in which the entire crew died as the ship broke up on its descent, just before its scheduled landing in Florida. Despite the blow to the Israeli people's spirit caused by the tragedy of his death, Ramon's widespread popularity brought major publicity to the Israeli space program and increased public support for future Israeli space missions. However, adequate funding in the government's annual budget continues to be a major problem for ISA.

The Israeli space program received a boost to its international standing in June 2003, when Israel was accepted by the European Space Agency (ESA) as a participating member. This role allows Israel to take part in ESA space projects and to make proposals for future joint projects with other ESA member states. Israeli space scientists have created an important niche for themselves in the worldwide space community by concentrating their research in the fields of remote sensing and the development of miniature satellites.

Launch Site Description

The Palmachim launch facility is located along the Mediterranean coast near the town of Yavne, which is south of Tel Aviv. Although the existence of the site officially remains a secret, portions are reported to be visible from the coast highway.

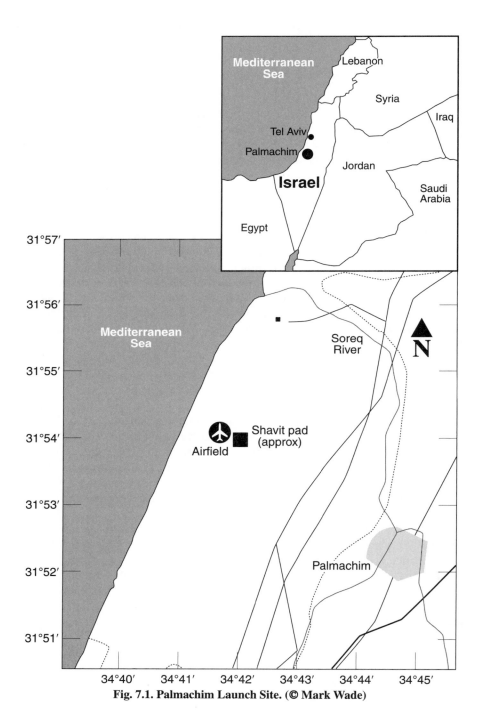

Fig. 7.1. Palmachim Launch Site. (© Mark Wade)

Palmachim was originally used for missile tests by the Israeli military prior to its use as the launch site for Shavit rockets in the 1980s. Its location allows for flight trajectories over water, which prevents rocket overflights in the airspace of neighboring nations. Shavit rockets are launched into retrograde orbits, on a westward trajectory over the Mediterranean Sea, passing over the Straits of Gibraltar at the western end of the Mediterranean. Launch capability is reported to range from a minimum inclination of 142.0 deg to a maximum inclination of 144.0 deg. As a result of the necessity of launching rockets westward from Palmachim, Israel has turned to other nations, including France and Russia, for the launch of its commercial satellites. Shavit vehicles are reportedly launched from a mobile transporter/erector/launcher, possibly after being assembled horizontally, separately from the payload.

While much information about Palmachim remains classified, reports have stated that the site has one runway, seven hangars, and several manufacturing facilities within its security perimeter. The actual launch site and assembly building are located at the southern end of the facility.

Acronyms

ESA   European Space Agency
ISA   Israel Space Agency
NCSR   National Committee for Space Research

Point of Contact

Israel Space Agency
P.O. Box 17185
26a Chaim Levanon Street
Ramat-Aviv, 61171 Tel Aviv, Israel
Voice: (+972) 3 216 852
Fax: (+972) 3 642 2298

References

S. J. Isakowitz, J. P. Hopkins Jr., and J. B. Hopkins, *International Reference Guide to Space Launch Systems,* 4th ed. (AIAA, Reston, VA, 2004).

Israel Space Agency (unofficial site), last modified 9 May 1999, <http://www.geocities.com/CapeCanaveral/5150/> (16 February 2005).

"Palmachim/Yavne—Israel Airfields," *GlobalSecurity.org*, last modified 5 December 2004, <http://www.globalsecurity.org/military/world/israel/palmachim.htm> (16 February 2005).

"Shavit," *Federation of American Scientists: FAS Space Policy Project—World Space Guide*, 2004, <http://www.fas.org/spp/guide/israel/launch/> (16 February 2005).

"Space Rocket Launch Sites Around the World, *Space Today Online*, 2004, <http://www.spacetoday.org/Rockets/Spaceports/LaunchSites.html#Palmachim> (16 February 2005).

M. Wade, ed., "Palmachim," *Encyclopedia Astronautica*, 2003, <http://www.astronautix.com/sites/palachim.htm> (16 February 2005).

8 Japan

History

Japan was one of the first nations to begin an ongoing program of space studies. That program originated in 1955, when a group of scientists and engineers from the University of Tokyo launched a 23 cm, solid-propulsion sounding rocket, dubbed the "pencil rocket," to study the atmosphere. The launch was initiated by Professor Hideo Itokawa and a team of volunteers as an engineering research project. This first venture into space exploration followed the resumption of aeronautical research in the aftermath of post–World War II restrictions that had been placed on Japan by the Allies. Japan then quickly made the commitment to participate in the events of International Geophysical Year (IGY, July 1957–December 1958) by launching a domestic sounding rocket. Early launches were from Akita, a suborbital launch site abandoned in 1962 for Kagoshima Space Center (KSC). Akita launched sounding rockets from September 1956 until May 1962.

Increased interest on the part of space scientists throughout Japan led to the founding of the Institute of Space and Aeronautical Sciences (ISAS) as part of the University of Tokyo in 1964. It was to serve as a joint research institute for observing space using space carrier vehicles. Studies were conducted during the late 1960s to develop a comprehensive government policy for future onboard scientific satellite projects and, as a result, the National Space Development Agency (NASDA) was founded in 1969. NASDA was assigned responsibility for developing launch vehicles and satellites as well as the facilities and equipment required by the Japanese space program, while ISAS was to continue its scientific research. Coordination of Japan's space efforts was to be overseen by the Space Activities Commission.

At its founding NASDA was charged with the mission of placing satellites into geostationary orbits. Judging it impractical to initiate an entirely domestic space effort, NASDA decided to import American technology—something the United States agreed to in 1969. Development began on the N series of rockets, which was based on technology transferred from the Delta series liquid-fuel rockets. The U.S. technology transfer enabled NASDA to acquire liquid-fuel rocket technology in a relatively short time.

Meanwhile, in 1970 Japan became only the fourth nation to launch its own satellite. This was the *Ohsumi*, launched by ISAS via a *Lambda 4S-5* solid-propellant rocket from KSC, now known as Uchinoura Space Center (USC). Development of the N series rockets proceeded, and in 1977 Japan successfully orbited the geostationary engineering testing satellite *Kiku-2,* launched by NASDA with the third *N-I* liquid-propellant rocket from Tanegashima Space Center (TNSC). NASDA was, however, still far from obtaining the lift capability required to deploy heavier satellites, forcing the agency to commission two satellite launches to the United States before the next-generation *N-II* vehicle was operational in 1981.

In 1981, ISAS was renamed the Institute of Space and Astronautical Science and was established as a joint research organization for Japanese universities. During the 1980s and 1990s, Japan's space program tested and flew a variety of vehicles, with the primary emphasis being scientific exploration. The *Challenger* space shuttle accident in January 1986 spurred the Japanese to seek greater autonomy

57

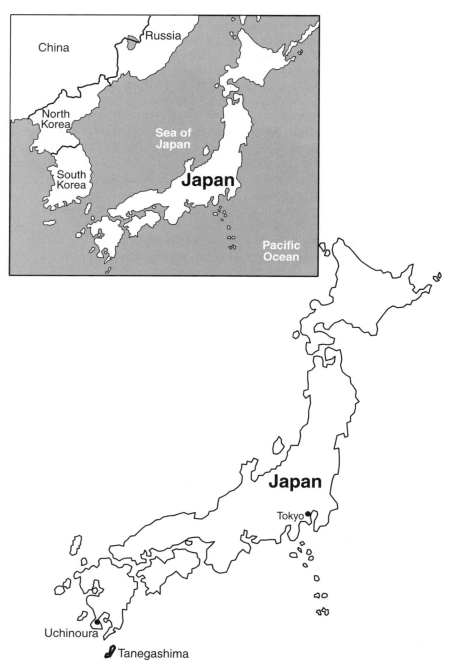

Fig. 8.1. Location of Uchinoura and Tanegashima launch sites within Japan.

from the United States for their space efforts. To achieve this, they sought to create an indigenous space transportation system and develop a highly reliable, more powerful launch vehicle that would ultimately make the launch of space payloads economically viable. The Japanese space program achieved many successes, operating with a relatively small budget compared with those of other space-faring nations, and it proceeded with a development plan for the *H-I* vehicle. NASDA enjoyed a high success rate with its *H-I* rocket and conducted nine *H-I* launches through 1992.

Japan's efforts toward manufacturing its own high-powered launch vehicle were next centered on the development of the *H-II* rocket, which was planned to lift a two-ton payload 36,000 km above Earth. During the early 1990s, Japan had hopes of entering the commercial launch market with the *H-II*, but it had to give them up in 1999 as a result of excessive operating costs and a series of program failures. The Japanese were forced to abandon the entire *H-II* program and proceed with the development of the *H-IIA*, a modified, cheaper version of the *H-II* rocket. On August 29, 2001, following several postponed launch attempts, the *H-IIA* was successfully launched into orbit from TNSC with a payload of two satellites.

A major reorganization of Japan's space program was conducted in October 2003, when NASDA was merged with ISAS and the National Aerospace Laboratory of Japan (NAL) to become the Japan Aerospace Exploration Agency (JAXA). The primary responsibilities of JAXA are to

- promote space-science research in cooperation with universities
- conduct scientific research and development in the fields of space science and aeronautics
- develop satellites and launch vehicles
- launch and track spacecraft

JAXA oversees the operations of both of Japan's launch sites, USC and TNSC.

The successful launch of an astronaut into orbit by China in October 2003, shortly after the creation of JAXA, required Japan to rethink its space policy priorities. Given its longstanding rivalry with China regarding which nation exerts the most influence in Asia, Japan began to consider the launch of a Japanese astronaut. The growing threat of North Korea caused another change in space program priorities, which forced Japan to attempt to place its own spy satellite into orbit to better survey North Korean military activities. Two spy satellites were successfully launched in March 2003, but the Japanese space program suffered a setback in December 2003, when an *H-IIA* rocket launched from Tanegashima carrying two spy satellites was forced to self-destruct only 10 minutes into flight.

The program received a needed boost in February 2005 when a dual-purpose navigation/meteorological satellite was placed into orbit by another *H-IIA* rocket from the Tanegashima site. The success spurred newspaper reports that JAXA was planning to begin crewed spaceflights. Some reports stated that the Japanese even intended to establish a crewed base on the moon by 2020 and to launch a Japanese-manufactured space shuttle by 2025.

Uchinoura Space Center
• orbital • 31.25 deg north • 131.08 deg east

USC was established by ISAS in February 1962 and officially opened in December 1963. Originally known as the Kagoshima Space Center, the launch site was renamed in October 2003. The primary purpose of the facility was originally to launch atmospheric sounding rockets and meteorological rockets. Later the site was used for launching satellites into low Earth orbit; the first six Japanese satellites were launched from Uchinoura. The main purpose of USC is now to launch scientific payloads, such as astronomical survey satellites and planetary probes.

ISAS selected a site consisting of several uninhabited hills facing the Pacific Ocean near Uchinoura in the Kagoshima Prefecture on the island of Kyushu, the southwesternmost of Japan's four main islands. Because its launch facilities are located on eight leveled hilltops at different heights covering 0.71 square km, Uchinoura has a unique appearance among the world's launch sites. Although situated about 320 m above sea level on the Nagatsubo plateau, USC is easily reachable by land as well as by port. Direct access is not a problem for potential users of the site.

By 1965, Uchinoura was equipped to launch the solid-fuel Kappa and Lambda sounding rockets, and it was capable of launching a small satellite using a Lambda-class, four-stage rocket by 1966. In February 1970, a Lambda-class launcher successfully placed Japan's first satellite into orbit. More recently, the M-V three-stage orbital rocket was successfully launched there starting in 1997. Since the founding of Uchinoura, hundreds of suborbital launches and more than two dozen orbital launches have taken place at the site.

Mission Capability

Space launches from Uchinoura are restricted to 45-day launch windows during January–February and August–September time frames because of range safety procedures. The limited launch windows are the result of successful pressure by the local fishing industry, which argued that launches pose too great a disturbance and danger to fishermen. The weather at USC is generally warmer and has greater precipitation than weather in the rest of Japan, a result of the influence of the Japan Current, which creates a subtropical climate on Kyushsu.

Launch Complexes

The KS Center (sometimes referred to as the Lambda launch complex), about 277 m above sea level, includes several mobile launchpads and a room to handle explosives.

The larger Mu launch complex is about 220 m above sea level and contains the Mu service tower, assembly shop, launch control blockhouse (about 80 m away from the launchpad), satellite test shop, dynamic-balance test shop, and propellant store. The Mu launch vehicle is assembled on its launcher inside the service tower in a vertical position. The M-V, like all other Mu launch vehicles, is attached to a vertical launcher rail that can be set at a specific elevation angle.

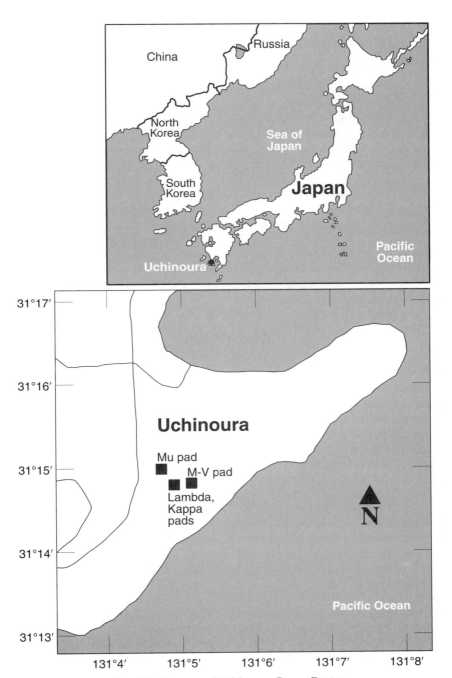

Fig. 8.2. Site map of Uchinoura Space Center.

Once the vehicle and spacecraft assembly is complete, the service tower is oriented toward the launch direction. The launcher with the Mu launch vehicle is drawn out of the service tower and tilted to the appropriate launch elevation angle (75 to 80 deg). In the 1980s, the Mu launch complex was upgraded to support the more powerful *Mu-3SII* vehicle. This launch vehicle was used to launch the Halley's Comet probes Sakigake (Pioneer) and Suisei (Comet) in January and August 1985. In the 1990s, the complex's facilities were upgraded for the M-V, with the first launch on 12 February 1997.

The M-V launch complex consists of a launchpad, assembly and launch tower, assembly building, and service building for rocket payloads.

Acronyms

ISAS	Institute of Space and Aeronatical Sciences (after 1981, Institute of Space and Astronautical Sciences)
JAXA	Japan Aerospace Exploration Agency
NAL	National Aerospace Laboratory of Japan
NASDA	National Space Development Agency
USC	Uchinoura Space Center

Point of Contact

Uchinoura Space Center
1791-13 Minamikata
Uchinoura-cho
Kimotsuki-gun
Kagoshima 893-1402
Japan
Voice: 81-994-31-6978
Fax: 81-994-67-3811

References

S. J. Isakowitz, J. P. Hopkins Jr., and J. B. Hopkins, *International Reference Guide to Space Launch Systems,* 4th ed. (AIAA, Reston, VA, 2005).

Japan Aerospace Exploration Agency, 2003, <http://www.jaxa.jp/index_e.html> (29 March 2005).

"Japanese Space Activities," *World Space Guide,* <http://www.fas.org/spp/guide/japan/index.html> (22 September 2005).

M. Wade, ed., "Kagoshima," *Encyclopedia Astronautica,* last modification 9 August 2003, <http://www.astronautix.com/sites/kagshima.htm> (29 March 2005).

Tanegashima Space Center
orbital • 30.4 deg north • 130.97 deg east

From the time Japan started to build a space program in the 1950s, plans were made to gradually develop an independent launch capability for geostationary satellites. Beginning in 1964, facilities on Niijima Island in Tokyo Bay were used to develop sounding-rocket technology. To launch satellites, Japan had to replace the Niijima Test Center with new launch facilities on the southeastern tip of Tanegashima Island (100 km south of USC and about 1000 km southwest of Tokyo).

The Tanegashima Space Center comprises the Osaki Launch Complex (a deactivated space launch facility), the Yoshinobu Launch Complex (an operational *H-IIA* facility), and the Takesaki Launch Complex (an active sounding rocket facility). The Osaki Launch Complex was established to operate the *N-I* vehicle, which was based on the Thor-Delta first stage and a Japanese engine (LE-3) second stage. On 23 February 1977, an *N-I* vehicle placed the *Kiku-2* satellite in geostationary orbit (making Japan, at the time, the only country besides the United States and the U.S.S.R. to achieve geosynchronous orbit). The Osaki Launch Complex was upgraded to accommodate the improved *N-II* launch vehicle, which launched a total of eight satellites between 1981 and 1987, including Japan's first sun-synchronous satellite (the *Momo-1* for remote sensing observations).

Mission Capability

Space launches from Tanegashima are restricted to two-month launch windows during January–February and August–September time frames because of range safety procedures. The limited launch windows are the result of successful pressure by the local fishing industry, which argued that for Tanegashima, as for Uchinoura, the launches pose too great a disturbance and danger to fishermen. Tanegashima has a relatively mild, temperate climate.

Two additional 60-day periods (June through July and November through December) have recently been made available for high-priority launches.

Facilities

TNSC is JAXA's largest facility. Located on Tanegashima Island, 115 km south of Kyushu, this 8.6-million-square-meter complex plays a central role in prelaunch countdown and postlaunch tracking operations. On-site facilities include the Masuda Tracking and Communication Station, the Nogi Radar Station, the Uchugaoka Radar Station (an L-band radar system near the Takesaki Range Control Center), and optical observation facilities to the west. Also present are related developmental facilities for firing liquid- and solid-fuel rocket engines.

The **Yoshinobu Second Spacecraft Testing and Assembling Building (#2STA),** in addition to unpacking the two-ton satellites to be launched with the *H-IIA*, is used for conducting various experiments such as radio wave characteristics and compatibility tests. After testing, a satellite is transported to the neighboring spacecraft and fairing-assembly building, for loading of propellant and final joining

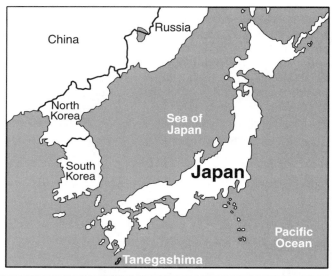

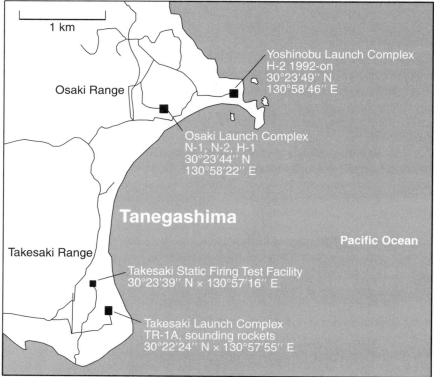

Fig. 8.3. Site map of Tanegashima Space Center.

with the fairing. Yoshinobu Firing Test Stand is used as the site for acceptance test firing of the *H-IIA/H-II* launch vehicle's first-stage LE-7 and the *H-IIA*'s LE-7A engines. Located next to the Yoshinobu Launch Complex, it includes storage and supply facilities for propellants (liquid hydrogen, liquid oxygen, helium, and nitrogen) and supply facilities for water and electricity.

Takesaki Range Control Center (RCC) collects and analyzes information, issues commands, makes adjustments, and monitors the prelaunch operations for rockets and satellites at Tanegashima. It is also concerned with ground safety, launching, and tracking. It serves as a control center for communications and coordination with tracking stations as well as downrange stations around the world. Three radar stations are on Tanegashima, at Nogi and Uchugaoka, plus an L-band radar on TNSC near the RCC.

The **Masuda Tracking and Data Communications Station** is located in the middle of Tanegashima Island. In addition to satellite tracking and control, the station is involved in the inspection and confirmation of the function of satellites on rockets at the Osaki Rocket Launch Complex prior to liftoff. It features the same equipment as JAXA's other tracking and communication stations.

Katsuura Tracking and Communication Station, located in the southeastern part of Kyushu Island's Boso peninsula, was constructed for satellite tracking and control. It has advanced equipment, including range-and-range rate tracking, telemetry reception, and command transmission systems.

The Okinawa Tracking and Communication System, located on the Japanese island of Okinawa, is JAXA's southernmost tracking facility. Its primary purpose is the tracking and control of satellites. The station has several facilities, including two parabolic antennas and a tracking and control building, that enable ground personnel to send commands to satellites and to determine satellites' altitude, location, and instrument functions.

The Takesaki Static Firing Test Facility is currently used for the firing of the solid rocket boosters for the *H-IIA* (SRB-As). The facility is equipped with a camera room, a measuring room, and a test stand with shelter.

Launch Complexes

Osaki Launch Complex

The Osaki Launch Complex, a midsize *J-I* launch complex, employs a 6.4 m high and 12 m wide launch deck (with a 170,000 kg mass) with a cantilevered steel structure. Two umbilical masts (43 m and 49 m) support the feed pipes and conduits for liquid oxygen/hydrogen, electrical power, air conditioning, etc. Prior to launch, the mobile service tower moves approximately 100 m away from the launchpad.

The first two stages of the launch vehicle are processed at vehicle assembly buildings 1 and 2 prior to erection on the pad. A solid motor test building is used for solid motors and pyrotechnics. A nondestructive test facility accommodates ultrasonic and x-ray inspection of upper stages and apogee kick motors. The spin test building accommodates integration of upper stages with its spin table. The spacecraft test and assembly building processes payloads in a class-100,000 clean room in preparation for mating with the upper stage, which is performed in the third stage and spacecraft assembly building. The Osaki blockhouse, a semiunderground, explosionproof building, is located 170 m away from the pad.

The Osaki Launch Complex was upgraded to handle the *H-I* launch vehicle by adding an additional cryogenic propellant facility and another, taller, service tower to the existing N1/2 complex. The first *H-I* test launch was successfully performed in August 1986, and the last launch took place on February 11, 1992.

This site was renovated for *J-I* vehicles, with the first *J-I* launch in February 1996 (the only J-I launch to date).

*Yoshinobu Launch (*H-IIA/H-II*) Complex*

The Yoshinobu Complex is the launch base for the *H-IIA* rocket. Its main facilities include the vehicle assembly building (VAB) for rocket assembly, preparation and checkout; two movable launchers (ML); two launchpads (LP-1 and LP-2); the pad service tower (PST) for final checks of launch vehicles on LP-1, propellant storage, and supply facilities; and the launch control building (blockhouse) for coordinating and issuing instructions for on-site launch operations. Completed in 1993, this complex was used for launching *H-II* rockets. In 1999, it was refurbished for *H-IIA* launches.

Takesaki Launch Complex

The Takesaki Launch Complex, located about 2.5 km south of the Osaki Range, is used for small launch vehicles. The site, which became operational in 1968, comprises two launchpads. From 1988 through 1989, it supported launches of the *TR-1* single-stage rocket that was utilized in development testing in support of Japan's *H-II* rocket. From 1991, this complex was used to launch the *TR-IA*, an improved *TR-I*, for space experiments. The last *TR-IA* launch was November 1998.

Acronyms

ML	mobile launcher
MST	mobile service tower
PST	pad service tower
TDAS	tracking and data acquisition station
TNSC	Tanegashima Space Center
VAB	vehicle assembly building

Fig. 8.4. *H-IIA* launch vehicle GTV, ground firing test at Osaki Range, Yoshinobu Launch Complex LP-I/PST . (JAXA)

Point of Contact

Tanegashima Space Center
Mazu, Kukinaga
Minamitane-machi, Kumage-gun
Kagoshima 891-3793
Japan
Voice: 81-9972-6-2111
Fax: 81-9972-4-4004

References

S. J. Isakowitz, J. P. Hopkins Jr., and J. B. Hopkins, *International Reference Guide to Space Launch Systems*, 4th ed. (AIAA, Reston, VA, 2005).

Japan Aerospace Exploration Agency, 2003, <http://www.jaxa.jp/index_e.html> (6 August 2005).

NASDA, "H-IIA Brief Description," *H-IIA Launch Vehicle User's Guide*, March 2000, <http://typhoon.aero.org/~fmd/IV.guides/> (6 August 2005).

NASDA, "Outline of the Tanegashima Space Center Range Facilities," 4th ed., November 2002, <http://h2a.jaxa.jp/documents/f4/tnsc_e.pdf> (6 August 2005).

M. Wade, ed., "Tanegashima," *Encyclopedia Astronautica*, last modification 30 March 2005, <http://www.astronautix.com/sites/tanshima.htm> (6 August 2005).

"World Space Guide," <http://www.fas.org/spp/guide/japan/facility/tanegashima.htm> (6 August 2005).

9 Norway

Andøya Rocket Range
• suborbital • 69.3 deg north • 16 deg east

Overview

The Andøya Rocket Range (ARR), located in northern Norway on Andøya Island, has responsibility for all aspects of the Norwegian space program. ARR was authorized by the Norwegian Ministry of Defense in 1960 for launching small sounding rockets and scientific balloons to conduct research in the atmosphere and ionosphere. It became operational in 1962, with the launching of three American Nike-Cajun rockets. Andøya is currently managed by the Norwegian Space Centre (NSC), headquartered in Oslo. Additional support for the range is provided by the European Space Agency (ESA). ESA funding stems from a cooperative agreement between the NSC and the ESA that allows for use of Andøya by some ESA member states. The facility is also utilized by the Norwegian Air Force and is available to other customers on a cost basis.

History

Launches of sounding rockets began when facilities at Andøya were in the early stages of development, and the range was subsequently occupied only during launch activities for its first few years. Although Norway was not a member of the European Space Research Organization (ESRO), founded in 1962, ARR was selected to host its launch operations, because ESRO's own facilities had not yet been completed. ESRO and the Norwegian government reached an agreement in 1965 to begin launching the French Centaure and Dragon rockets. The first six ESRO rockets were launched from Andøya in early 1966, and four rockets were also launched that year on behalf of the French space agency, the Centre National d'Études Spatiales (CNES). In September 1966, ESRO-sponsored launches were moved to the newly opened Esrange launch site at Kiruna, Sweden. (See chapter 12 for more about Esrange.) Despite the completion of Esrange, a diverse group of international customers, including NASA, continued to use the ARR.

By the end of 1972, after 10 years of operation, Andøya had launched 104 vehicles; 102 of the launches were successful. By 2004, more than 800 sounding rockets had been launched from ARR, ranging in size from the 1.52-m Dart to the larger, four-stage *Black Brant XII*. ARR has operated an associated launch site since 1997 at Ny-Alesund on the Arctic archipelago of Svalbard to provide additional launch capabilities.

Launch Site Description

ARR is located some six km from the town of Andenes. The range has eight launchpads, including a universal ramp capable of launching rockets weighing up to 20 metric tons. The largest launch facility has a load restriction of 20,000 kg. ARR customers have access to the 5000 m runway at nearby Andøya Air Force Base, courtesy of an agreement with the Royal Norwegian Air Force. Andenes has air connections to the rest of Norway and is also served by the port of

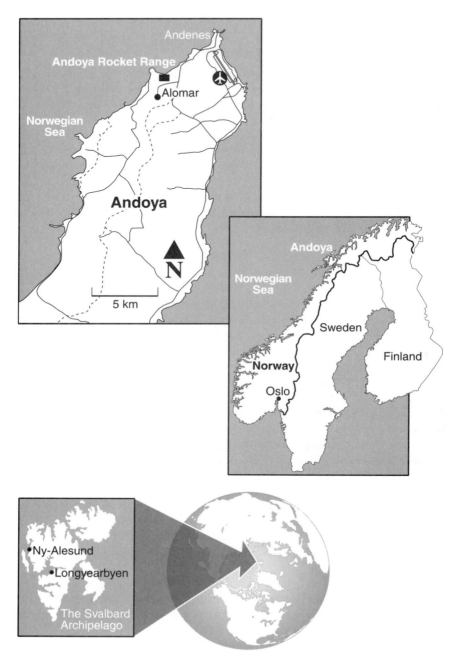

Fig. 9.1. Location of the Andøya Rocket Range in Norway and the launch site on the Arctic archipelago of Svalbard.

Risoyhamn, located about 50 km south. A full range of services is provided by ARR for launch, operations, recovery, and ground instrumentation support.

ARR has a huge impact area that offers a large range of rocket configurations and optional launch directions. The widespread impact area also affords customers the option of launching without guidance or flight termination systems, which enables payload increases and saves costs. Payload recovery service from the Norwegian Sea by ship or helicopter is provided by ARR, with no restrictions on payload size or weight. Clean-room facilities and payload-cleaning equipment are available for payload refurbishment and offer a rapid turnaround time if needed. In spite of ARR's high latitude, the climate is generally mild, even in midwinter, because the site's weather is moderated by the Gulf Stream. The area's climate also provides good conditions for balloon launch operations, and more than 450 balloons have been launched from ARR since its founding.

Associated Support Facilities

Svalbard Launch Site

The launch site on the Arctic archipelago of Svalbard (79 deg N) provides unique conditions for rocket studies of the dayside polar cusp, cleft, and cap, which can be conducted using smaller, less expensive sounding rockets. Svalbard's location is ideal for scientific exploration of the dayside aurora and the magnetospheric boundary layer. Sounding rockets can be launched directly into the polar cap and also along and perpendicular to Earth's magnetic field. Svalbard offers both a universal launcher and a payload integration facility.

ALOMAR Research Station

ARR recently established a new international research facility, ALOMAR (Arctic Lidar Observatory for Middle Atmosphere Research), which provides scientists opportunities for year-round studies of the Arctic middle atmosphere. ALOMAR has responsibility for all of the ARR's ground-based instrumentation. The facility consists of an observatory with optical instruments and several nearby ground-based radar sites. The observatory is located close to the ARR, on top of the 379-m Mt. Ramnan on the island of Andøya. The concept for ALOMAR was implemented by the Institute for Atmospheric Physics, University of Rostock, Germany, and the Norwegian Defense Research Establishment.

Acronyms

ALOMAR	Arctic Lidar Observatory for Middle Atmosphere Research
ARR	Andøya Rocket Range
CNES	Centre National d'Études Spatiales
ESA	European Space Agency
ESRO	European Space Research Organization
NSC	Norwegian Space Centre

Points of Contact

Norsk Romsenter (Norwegian Space Centre)
Hoffsveien 65A
P.O. Box 85 Smestad
N-0309 Oslo 9
Norway
Voice: +47 2 52 38 00
Fax: +47 2 23 97

Andøya Rocket Range
P.O. Box 54
N-8483 Andenes
Norway
Voice: +47 76 14 44 00
Fax: +47 76 14 44 01

Hallstein Thommassen
Andøya Rocket Range
P.O. Box 54
N-8480 Andenes
Norway
Voice: +47 76 14 16 44

References

Andøya Rocket Range, 2004, <http://www.rocketrange.no/arr/> (25 March 2005).

T. Guldvog, ed., *Norwegian Space Centre*, 2003, <http://www.spacecentre.no/> (25 March 2005).

"Norway's Andøya Rocket Range," *Space Today Online*, 2003, <http://www.spacetoday.org/Rockets/Spaceports/Norway.html> (25 March 2005).

J.-J. Serra, "Andøya," *Rockets in Europe,* <http://www.univ-perp.fr/fuseurop/andoya_e.htm> (25 March 2005).

M. Wade, ed., "Andoya," *Encyclopedia Astronautica*, last modification 9 August 2003, <http://www.astronautix.com/sites/andoya.htm> (25 March 2005).

10 Pakistan

SUPARCO Flight Test Range
• suborbital • 25.2 deg north • 66.75 deg east

Pakistan initiated its space program on 7 June 1962, with the successful launch of *Rehbar 1*, a Nike-Cajun sounding rocket. The mission's goal was to obtain meteorological data. Only four days later, *Rehbar 2* was launched. These launches took place one year after the formation of the Space and Upper Atmosphere Research Commission (SUPARCO), Pakistan's national space agency. SUPARCO was placed under the jurisdiction of the Pakistan Atomic Energy Commission (PAEC) and given responsibility for formulating a program to study the upper atmosphere. Shortly after SUPARCO was created, a sounding rocket launching facility was constructed at Sonmiani, some 50 km northwest of Karachi, which became the site for the Rehbar launchings and is still in use today. SUPARCO's meteorological sounding rocket program maintained a steady launch schedule through the remainder of the 1960s, which enabled the agency to carry out its upper-atmosphere research. During the 1970s, SUPARCO constructed ground facilities for tracking rockets and satellites and acquiring telemetry and other scientific data from weather, remote sensing, and navigational satellites.

By 1980, SUPARCO had begun the planning for an indigenous Pakistani satellite. To acquire experimental knowledge of satellites in low Earth orbit, it established two telemetry, tracking, and command ground stations at Karachi and Lahore. This experience was then utilized for the design and development of Pakistan's first satellite, *Badar-1*, which was successfully launched aboard a Long March rocket in July 1990 from a Chinese launch site. Pakistan's second satellite, *Badar-2*, was launched from the Baikonur Cosmodrome in 2001 using a Russian *Zenit-2* rocket as the launch vehicle. Another SUPARCO achievement was the construction of a satellite ground receiving station near Islamabad, which enabled Pakistani scientists to receive Earth-imaging data transmitted from U.S. and French satellites. In addition to space research, current SUPARCO objectives include "promoting the peaceful applications of space science and technology" to advance Pakistani socioeconomic development.

Launch Site Description
SUPARCO operates the national sounding rocket Flight Test Range (FTR). Although orbital launch operations are not available, small- to medium-class sounding rocket launches for scientific research can be arranged by contacting SUPARCO. The FTR, located at Sonmiani Beach in the district of Labella, covers an area of about 200 ha (500 acres). It was enlarged and modernized in the 1990s, and although announcements have periodically been made that the range will be upgraded to accommodate larger-class sounding rockets, no specific plans have been formalized. Typical sounding rockets launched from Sonmiani include the Centaure and Dragon series. More than 200 sounding rockets have been launched to an altitude range of 20–550 km from Sonmiani since the range was established in 1962. The typical lift capability is a 60-kg payload to an altitude of 440 km. Launch corridor azimuths at Sonmiani are limited from 220 to 310 deg.

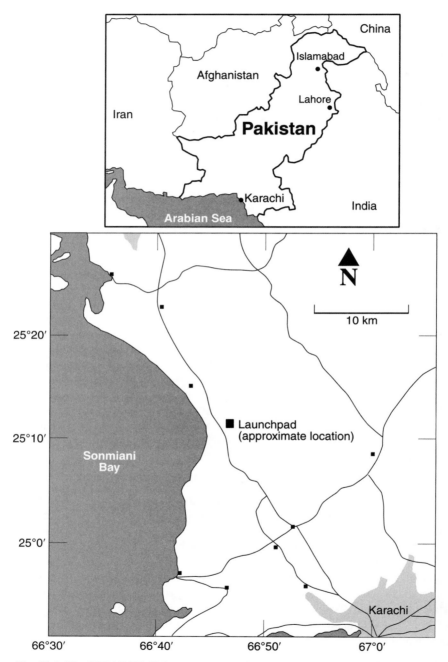

Fig. 10.1. The SUPARCO flight test range at Sonmiani. (Site map © Mark Wade)

The preferred launch season is from October to March, the months with the clearest weather conditions. The range is equipped with the necessary facilities for various experiments, including rocket launchers, rocket assembly workshop, payload preparation area, high-speed tracking radar, control room, telemetry station, flight communication equipment, optical cameras, and maintenance workshop. While the FTR is primarily used by Pakistan's civilian space program, it has also been used by the Pakistani National Defence Complex, also under the jurisdiction of the PAEC, for the testing of solid-fuel ballistic missiles.

Fixed-rail launchers of various types exist on separate launchpads. They are supported by a mobile crane and mobile service platforms. Facilities at the FTR include an assembly shop (rocket and payload integration hall), an electronics maintenance shop, a 60-m wind tower, and other meteorological support equipment. Payload accommodations are not available at the FTR. Port facilities are available in Karachi.

Associated Support Facilities

SUPARCO's instrumentation laboratories in Karachi provide the necessary instrumentation for data sensing, transmission, reception, and processing, as well as for the tracking of sounding rockets for upper-atmosphere research. The laboratories are equipped with the facilities to design and fabricate electronic devices used in rockets, satellites, and ground control equipment. They also contain telemetry systems, infrared instrumentation, vibration systems, digital circuit designs, microprocessor-based systems, and high-speed computer systems to calibrate and test such equipment.

Acronyms

FTR	Flight Test Range
PAEC	Pakistan Atomic Energy Commission
SUPARCO	Pakistan Space and Upper Atmosphere Research Commission

Point of Contact

Pakistan Space and Upper Atmosphere Research
Commission (SUPARCO)
Sector 28, Gulzar-e-Hijri
P.O. Box 8402
Karachi 75270
Pakistan
Voice: (92-21) 8144667-74, 8144923-927
Fax: (92-21) 8144928, 8144941
E-mail: suparco@digicom.net.pk

References

"Pakistan Derives Its First 'Hatf' Missiles from Foreign Space Rockets," *Wisconsin Project on Nuclear Arms Control*, 2005, <http://www.wisconsinproject.org/countries/pakistan/hatf.html> (25 March 2005).

"Pakistan Launch Vehicles," *FAS Space Policy Project—World Space Guide*, last modified 20 March 2000, <http://www.fas.org/spp/guide/pakistan/launch/index.html> (25 March 2005).

Pakistan Space and Upper Atmosphere Research Commission, 2002, <http://www.suparco.gov.pk/> (25 March 2005).

"Resources on India & Pakistan: Chronology of Pakistani Missile Development," *Monterey Institute of International Studies*, 2002, <http://cns.miis.edu/research/india/pakchron.htm> (25 March 2005).

M. Wade, ed., "Sonmiani," *Encyclopedia Astronautica*, <http://www.astronautix.com/sites/sonmiani.htm> (11 April 2005).

11 Russia and Kazakhstan

Overview

The space age began with the launch of the satellite *Sputnik 1* by the Soviet Union on 4 October 1957. This single event was largely responsible for the rapid acceleration of the exploration of outer space that occurred over the next few years. Without Sputnik, the United States, which launched its own Explorer satellite in January 1958, would never have made such a tremendous effort to compete with the Soviets for space superiority. While space exploration certainly was already proceeding at its own pace prior to Sputnik, the launch of the first satellite by the leading Cold War adversary of the United States guaranteed a rapid start to the "space race."

At the close of World War II, the Soviet Union embarked on an all-out program to develop rocketry. Like the Americans, the Soviets utilized the skills of captured German scientists who had worked with Nazi missile programs. With the full backing of the Communist government, the Soviets made rapid advances in their rocket program. Simultaneously, they made great strides toward developing an intercontinental ballistic missile (ICBM), which was vital to Soviet defenses because they had no long-range bomber fleet.

The Soviets made several public pronouncements that they planned to orbit a satellite during International Geophysical Year (July 1957–December 1958), but most Western observers wrote off these announcements as propaganda ploys. When the R-7 rocket, the world's first ICBM, successfully flew in the summer of 1957, the Soviets realized that they had a missile that could also serve as a launch vehicle to boost a satellite into low Earth orbit. Soviet Premier Nikita Khrushchev then gave his approval for the launch of Sputnik in October, and the Soviet space effort proceeded quickly under the guidance of chief designer Sergei Korolev.

Virtually all of the early space firsts were subsequently achieved by the Soviets in rapid succession. The first living creature in space, the dog Laika, was launched aboard *Sputnik 2* in November 1957; *Luna 2* became the first human-made object to reach the moon in 1959; the first human to orbit Earth, cosmonaut Yuri Gagarin, was launched from Baikonur in April 1961; the first two-man mission was flown in 1962; the first woman in space, Valentina Tereshkova, flew aboard the *Vostok-6* in June 1963; and the first space walk was performed by cosmonaut Alexei Leonov in March 1965.

When U.S. President John F. Kennedy pledged that the United States would land a man on the moon by the end of the 1960s, the U.S. space program was lagging behind the Soviet Union's. By the mid 1960s, however, the United States was rapidly catching up. By the end of the series of Gemini program flights in 1966, the Americans had passed the Soviets in some aspects of spaceflight, including docking techniques, which would be critical to any future lunar landing.

As the American Apollo program proceeded in the late 1960s, the Soviets fell farther behind in the race to land a person on the moon. Their effort was particularly hindered by their inability to develop a booster similar to America's powerful and massive *Saturn V.* When Apollo astronauts landed on the moon in July 1969,

Fig. 11.1. Russian launch sites.

the space race ended with an American triumph. Although the initial response of the Soviets was to deny that they had ever planned to land a cosmonaut on the moon, there is no doubt today that they had mounted a full-scale effort to achieve the first crewed lunar landing. The success of the Apollo program was a tremendous blow to the morale and prestige of the Soviet space program.

Despite the victory of the Americans in the race to the moon, the Soviet space effort continued apace throughout the 1970s and 1980s and achieved an impressive legacy. Its achievements included the orbiting of the Salyut space station, in 1974, and the Mir space station, launched in 1986. The crews of these space stations eventually achieved space duration records that are still unmatched in the history of spaceflight. Also, the Soyuz gradually obtained one of the highest launch reliability rates of the world's launch vehicles, becoming the workhorse of the Soviet fleet.

When the Soviet Union formally dissolved in 1991, the Russian Federation inherited most of the Soviet space program. Ukraine acquired its own space program when the newly independent nation inherited the space programs and facilities that were formerly located on Soviet territory. The Baikonur launch site, the location of the launches of *Sputnik 1* and Yuri Gagarin's *Vostok* spacecraft, found itself part of independent Kazakhstan, and Russia now leases Baikonur in order to retain this important site. The Soviet breakup caused a vastly reduced budget for the Russian space program, and funding is still far below what is needed to adequately maintain what was once the world's most impressive presence in space.

In 1993, Russia joined the International Space Station (ISS) effort, and in 1995 a U.S. space shuttle docked with space station Mir for the first time. Following the space shuttle *Columbia* disaster in February 2003 and the grounding of the shuttle fleet, Russia's Soyuz rocket became the only way to replace crews and ferry new supplies to the ISS.

The Russian Federation currently uses two principal launch sites, which are designated as cosmodromes (the Russian term for launch sites): Baikonur, which has been located in Kazakhstan since the breakup of the Soviet Union but is currently leased by Russia, and Plesetsk, a site primarily used for Russian military launches. In the first few years after 2000, Baikonur was the world's busiest launch site, despite severe cutbacks in the Russian space budget.

Russia has two other launch sites, the Kapustin Yar Cosmodrome and the Svobodny Cosmodrome. Kapustin Yar, also known as the Volgograd Station, is located in southwestern Russia near the city of Volgograd. It was the first missile-testing center in the Soviet Union. Beginning in 1947, the site was used by the Soviet military to test captured German V-2 rockets, and it was later used for testing ballistic missiles and for satellite launches using Kosmos rockets. While Kapustin Yar is still used occasionally for missile testing and launching sounding rockets, it is rarely utilized today for launching satellites, because Russian military satellite launchings were transferred to Plesetsk in the 1980s. Only one orbital satellite launch has taken place there since 1987.

Svobodny is Russia's newest cosmodrome, established by official decree in 1996. The site was founded after the breakup of the Soviet Union, when Russia realized that it needed another launch site. Baikonur was now located in Kazakh territory, and facilities at Plesetsk could not support large launch vehicles. Svobodny is a former Soviet missile base in Russia's Far East, but although the site was upgraded and declared operational for use by space launch vehicles in 1997, only four satellite launches have taken place there. Inadequate funding has stalled many of the planned uses for Svobodny, but Russia still hopes that the site's launchpads can be converted for future use by Proton and Angara launch vehicles.

Plesetsk Cosmodrome
• orbital • 40.5 deg east • 62.8 deg north

Overview

The Plesetsk Cosmodrome is Russia's northernmost cosmodrome, located in the Archangel region of the Russian Federation, near the towns of Plesetsk and Mirny. Plesetsk, once the world's most active launch site, is primarily used for launching military satellites into polar orbit. Founded in 1957, the same year the Soviet Union launched *Sputnik 1*, as an ICBM base, Plesetsk grew in tandem with the rapid developments of the space age. It was the world's first operational ICBM base; the launchpad for the R-7 ICBM was declared operational in December 1959.

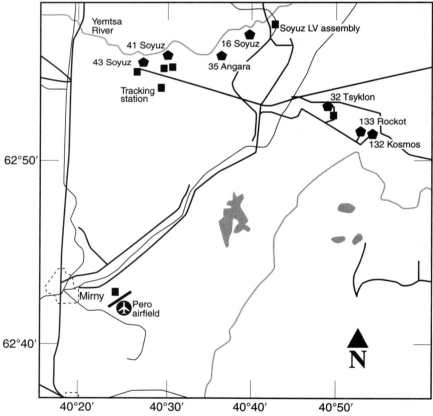

Fig. 11.2. The Plesetsk Cosmodrome. (site map © Mark Wade)

In the early 1960s, the Soviets decided that a launch site was needed to place satellites into polar and high-inclination orbits, which they were unable to do from the existing Baikonur launch site, and Plesetsk was officially designated a cosmodrome in January 1963. Living quarters for workers and support facilities were established at the nearby settlement of Mirny, which formally became a city in 1966. The first space launch from Plesetsk occurred in March 1966 with the orbital launch of a *Cosmos 112* satellite onboard a *Vostok-2* booster. Beginning in 1968, Plesetsk was the site for the testing of early models of mobile ICBMs.

Although the cosmodrome took its name from the nearby village of Plesetsk, which had already been established when the ICBM base was constructed, the launch site is also sometimes referred to as Mirny, because that town later became the principal base of support operations. Since its founding, Plesetsk has been the site for launches of communications, navigation, reconnaissance, meteorological, and other satellites, including many in the Molniya series. Despite the high level of launch activities at Plesetsk, the Soviets did not officially acknowledge the cosmodrome's existence until 1983. In recognition of the important role that Plesetsk has played in the Soviet/Russian space program, in 1994 the facility was given the title of "First State Testing Cosmodrome."

With the dissolution of the Soviet Union in 1991 and the establishment of the Russian Federation, launch activity suffered a downturn as a result of cutbacks in the Russian military budget. The number of launches decreased from a total of 47 in 1988 to only six by 1996. By 2002, more than 1900 satellites had been launched from Plesetsk, but Baikonur has overtaken Plesetsk as the site for most of Russia's launch campaigns. This largely results from the reduction of Russian military launches, which primarily occur at Plesetsk, and an increase in commercial GTO (geostationary transfer orbit) missions and ISS launches from Baikonur.

Launch Site Description

The Plesetsk Cosmodrome is an enormous launch complex, covering an area of approximately 1700 sq km. It is located some 800 km northeast of Moscow and about 200 km south of Archangel. The cosmodrome currently has six launch complexes with more than 11 launchpads capable of supporting a variety of launch vehicles, including Kosmos-3M, Soyuz, Molniya, *Tsyklon-3*, and Start, first test-flown at Plesetsk in 1993. A launch complex was planned for the Zenit rocket, but it was never completed. It is in the process of being reconfigured to accommodate future launches of the Angara launch vehicle.

The site's northern latitude restricts space missions to orbital inclinations between 63 and 83 deg. The launch corridors from Plesetsk allow the suborbital stages of boosters to fall back onto Russian territory.

The northerly location of Plesetsk has given the Russians extensive experience with cold-weather launches. Winter temperatures can drop to −38°C, and summer temperatures can reach as high as 33°C. Average rainfall is approximately 398 mm annually, and maximum ground winds are 18 km/h.

The first launch complex to be constructed at Plesetsk, Launch Complex (LC) 41, was completed in 1959, and by the next year LC-16 was also declared operational. All the launchpads at Plesetsk were initially used for test launches of Soviet ICBMs. After the launch of the first spacecraft in 1966, launch activity accelerated at a rapid pace. Despite the downturn in the frequency of Russian military satellite launches, Plesetsk is still utilized as a launch site, and Russia hopes that it will be used increasingly by commercial launch operators. The first commercial launch at Plesetsk took place in 1990, and such launches continue there regularly.

There are active launch complexes for the Kosmos 3-M, Tsyklon-3, Soyuz, Molniya, Start, and Rockot launch vehicles. A former Kosmos pad, LC-133, has been reconfigured for use by Rockot launch vehicles. The Russian Design Bureau of Transport Machinery (KBTM) oversees the launch complexes and ground infrastructure for Rockot, Kosmos, and Tsyklon rockets.

Plesetsk has two types of launchpads. With one type, the booster is assembled on the pad and surrounded by a service structure or mobile building until launch preparations are under way. (This procedure is similar to the way Titan launchpads were formerly used at Cape Canaveral.) With the other, a launch vehicle is first horizontally integrated and assembled with its payloads at another location. It is then transported by rail to the pad.

Plesetsk contains launch vehicle assembly buildings, telemetry and tracking stations, testing facilities, and equipment for spacecraft preparation and checkout. It also has the largest oxygen liquefaction plant in Europe. On-site accommodations are provided for workers in launch campaigns, with housing located 1–2 km away from launchpads. Real-time communications to any place in the world are available for Plesetsk customers. The launch complex has a well-developed transportation infrastructure, including an airfield. In 2005, plans were announced to extend the airfield's 2600 m runway and to modernize or replace other airport equipment. Currently, the airport accommodates TU-154 and Il-76 transport planes.

Mirny, a planned city with a population estimated (in 2004) at 80,000, is located about 36 km from Plesetsk. It is connected to the launch site by both road and rail. In addition to living quarters for workers from Plesetsk, Mirny contains a hotel and recreational facilities for visitors.

Acronyms

ICBM	intercontinental ballistic missile
ISS	International Space Station
KBTM	Russian Design Bureau of Transport Machinery
LC	launch complex

Points of Contact

Plesetsk Space Center
Mirniy-12, Arkhangelskaya Oblast,
164170 Russia
Voice: +7 (095) 330-9190
Fax: +7 (095) 330-9190

Design Bureau of Transport Machinery (KBTM)
101 Vernadsky Prospect, Bldg. 2
Moscow 117415
Russia
Voice: +7 (095) 433-3239
Fax: +7 (095) 433-1548
E-mail: kbtmto@dol.ru

Kosmos

Cosmos International Satellitenstart GmbH
Universitätsallee 29, 28359 Bremen
Voice: +49 (0) 421 2020-8
Fax: +49 (0) 421 2020-700
E-Mail: cosmos@fuchs-gruppe.com

Proton

International Launch Services
1660 International Drive
Suite 800
McLean, Virginia 22102
U.S.A.
Voice: (571) 633-7400
Fax: (571) 633-7500

Rockot

Eurockot Launch Services GmbH
P.O. Box 28 61 46
D-28361 Bremen
Voice: +49 421 539-65 01
Fax: + 49 421 539-65 00
E-mail eurockot@astrium-space.com

Soyuz

Starsem
2, rue François Truffaut
91042 Evry Cedex
France

Voice: +33 1-69-87-01-10
Fax: +33 1-60-78-31-99

Zenit
United Start Corporation
2995 Airway Avenue
Costa Mesa, CA 92626
U.S.A.
Voice: (714) 755-7427
Fax: (714) 545-7676
Email: info@unitedstart.com

References

"Centers: Plesetsk," *RussianSpaceWeb*, <http://www.russianspaceweb.com/plesetsk.html> (31 March 2005).

I.-S. Chang, "Space Launch Vehicle Reliability," *Crosslink* **2** (1) (Winter 2000/2001).

Cosmodrome Plesetsk, <http://plesetsk.org/> (31 March 2005).

S. J. Isakowitz, J. P. Hopkins Jr., and J. B. Hopkins, *International Reference Guide to Space Launch Systems,* 4th ed. (AIAA, Reston, VA, 2005).

Jonathan's Space Report, <http://www.planet4589.org> (7 August 2005).

C. Lindborg, "Plesetsk Cosmodrome," *FAS Space Policy Project: World Space Guide*, <http://www.fas.org/spp/guide/russia/facility/plesetsk.htm> (31 March 2005).

"Plesetsk Cosmodrome," *Design Bureau of Transport Machinery (KBTM)*, 2001, <http://www.kbtm.ru/english/complexes/plesetsk.htm> (31 March 2005).

"Plesetsk Cosmodrome," *U.S. Centennial of Flight Commission*, 2003, <http://www.centennialofflight.gov/essay/Dictionary/PLESETSK/DI170.htm> (31 March 2005).

User's Manual for Eurockot Launches from Plesetsk Cosmodrome, Issue 4, Revision 1 (Eurockot GmbH, Bremen, Germany, 2005).

M. Wade, ed., "Plesetsk," *Encyclopedia Astronautica*, last modification 30 March 2005, <http://www.astronautix.com/sites/plesetsk.htm> (31 March 2005).

Baikonur Cosmodrome
• orbital • 45.60 deg north • 63.40 deg east

Overview
The Baikonur Cosmodrome has been the site of some of the most historic launches of the space age, dating back to the launch of the world's first artificial satellite, *Sputnik 1*, in October 1957 and the first spaceflight of a human, cosmonaut Yuri Gagarin, in April 1961. Baikonur is the world's oldest space launch complex, but it has maintained a consistent presence throughout the history of space exploration and remains one of the world's busiest spaceports.

Baikonur received renewed attention and significance beginning in the mid-1990s because of its importance in maintaining the operations of the ISS by means of Soyuz rockets launched from Baikonur. In addition to Soyuz launch vehicles, the Baikonur Cosmodrome can also launch Zenit, Proton, Rockot, Molniya, Tsyklon, Kosmos, and Dnepr rockets.

The Baikonur launch site was founded as a testing facility for the R-7 ICBM. An extremely remote location was selected, partially for reasons of secrecy and also because the huge area of land that the site encompassed could accommodate the R-7's radio guidance system. Construction began in 1955 and was completed by the end of 1956. The city of Leninsk was constructed about 48 km from the spaceport to serve as the living quarters for Baikonur workers. The cosmodrome became operational in May 1957, and in August the world's first successful ICBM flight took place from the launchpad of Area 1.

Less than six weeks later, Baikonur became the birthplace of the space age when *Sputnik 1* was launched into orbit from the same pad. Since the subsequent launch of Yuri Gagarin, this pad, now known as the "Gagarin pad," has been the site of more than 400 launches. A second pad capable of supporting crewed launches was completed in 1965.

The name "Baikonur" has been a source of confusion since the launch site was founded. To keep the site's location a secret from Western intelligence sources, the Soviet Union used the same coordinates for the site as the town of Baikonur, which is actually some 350 km to the northeast. The closest town to the launch site is actually Tyuratam, a railhead junction located south of the launch site, and U.S. intelligence agencies soon learned the site's true location from photographs taken by U-2 spy planes. Both the Soviets and American intelligence sources privately referred to the site as Tyuratam, although Soviet officials continued to use the name Baikonur in public, and the site's location remained an official state secret for many years.

After the 1991 breakup of the Soviet Union, the launch site's location was determined to be part of the newly independent nation of Kazakhstan, and the new Russian Federation found it necessary to negotiate a 20-year lease agreement with the Kazakh government in 1994 to keep control of the spaceport's operations. Though Kazakhstan officially changed the site name to Tyuratam in 1992, Russia

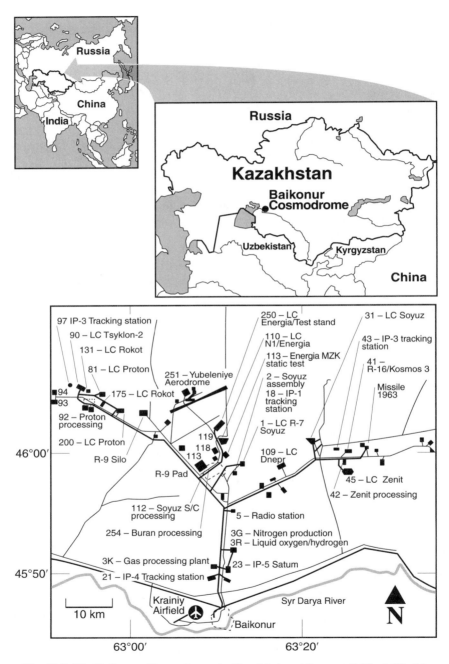

Fig. 11.3. The Baikonur Cosmodrome in Kazakhstan. (site map © Mark Wade)

and the rest of the international space community have continued to refer to the site as Baikonur. The city of Leninsk was officially renamed Baikonur in 1995.

All crewed Russian spaceflights originate at Baikonur, and since 1991 it has become the site of almost all Russian nonmilitary launches. Baikonur also serves as the site for most geostationary orbit, planetary, and lunar launches. In recent years, Baikonur has become a highly utilized site for commercial launches, and several companies now operate on-site facilities in support of commercial launch campaigns.

In March 2005, Kazakh officials announced plans to study the possibility of developing an autonomous launch site that would have the ability to launch small commercial satellites into low Earth orbit. The proposed launch complex would utilize MiG-31 fighters to boost rockets and their payloads to a high altitude. The rockets would then be fired from the MiGs to launch satellites into orbit. In June 2005, Russia signed an agreement with Kazakhstan to extend its lease of Baikonur through 2050.

Launch Site Description

Baikonur Cosmodrome is located about 2400 km southeast of Moscow on the arid central steppes of Kazakhstan. The enormous launch site covers an area of approximately 75 km from north to south and 90 km from east to west. The Cosmodrome has a harsh, dry climate with extremes of heat and cold. Temperatures can range up to 40–45°C in the summer, while lows in the winter sometimes reach 25–35° below 0°C. Baikonur has traditionally been divided into three areas, known as the "Center," "Left Flank," and "Right Flank." The cosmodrome has nine launch complexes with 15 operational launchpads; Baikonur has more than 50 launchpads altogether. The various launch vehicles launched from Baikonur include Soyuz, Molniya, Zenit, Proton, Rockot, Tsyklon, Kosmos, and Dnepr. Eleven assembly and checkout facilities are available for assembling and testing launch vehicles.

Additional launch support facilities include a medical center dedicated to spacecraft crews, a gas-turbine power plant, an oxygen-nitrogen plant, a dynamic testing stand, two airports, a water supply system, three fueling facilities, and an extensive highway and rail network. Access to international communications systems is provided through an existing cosmodrome network consisting of fax machines and telephones. Customers also have the option to rent channels from satellite operators, such as Eutelsat and Intelsat, for phone, fax, data, E-mail, Internet, and videoconference communication between Baikonur Cosmodrome and spacecraft control centers worldwide.

Because most Russian military launches no longer take place at Baikonur, much of the cosmodrome is under the jurisdiction of Rosaviacosmos, the Russian Aviation and Space Agency. Now that Baikonur is increasingly dedicated to civilian and commercial launches, several Western firms and joint business ventures conduct on-site commercial operations and have made considerable investments in the infrastructure of Baikonur. These businesses include Starsem, a multinational European-Russian firm that conducts commercial Soyuz launches; Kosmotras, a joint Russian-Ukrainian company that oversees launches of the Dnepr rocket; International Launch Services, a U.S.-based joint venture that conducts Proton launches; and the United Start Corporation, which launches Tsyklon rockets. In addition, the Ukrainian Yangel Yuzhnoye Design Bureau is responsible for Zenit launches.

Launch services at Baikonur are grouped according to individual launch vehicles. Following are descriptions of the launch complexes and associated support facilities for some of the rockets that are most frequently launched from Baikonur.

Proton

Launch Complex

Baikonur Cosmodrome is the only launch site with facilities capable of launching the Proton rocket. Krunichev State Research and Production Space Center utilizes the Proton launch vehicle for Russian planetary and geostationary spacecraft missions, and International Launch Services is responsible for most Proton commercial launches. Proton launch facilities are located on the Left Flank, the western side of the cosmodrome.

Four launchpads, two at Area 81 and two at Area 200, are dedicated to Proton, and two of these are configured to support commercial spacecraft launches. The two original Proton launchpads, numbers 23 and 24, are located at Area 81. Launchpad 23 is used for the Proton K/Block DM vehicle configuration. The first Proton launch from Pad 23 took place in 1967. Launchpad 24 is almost identical to 23 but has been modified to accommodate either a Proton K or Proton M with either a Block DM or a Breeze M. Pad 24, the oldest launch facility for the Proton, was the site of the first Proton launch in 1965. Both pads have undergone extensive renovations in recent years. Area 200 is the location of launchpads 39 and 40.

The launch area includes a launch structure with launchpad, including an underground vault; a launch vehicle mobile service tower (MST); and a bunker for observation of launches. The launch structure and vault house equipment supports the prelaunch processing of the launch vehicle. They provide electric, pneumatic, and hydraulic links between the ground system testing equipment and onboard launch vehicle hardware via the launch vehicle transit cables and pipes. The launch structure is designed to withstand launch vehicle first-stage engine plume impingement. The launchpad is intended for installing and erecting the Proton, and securing the launch vehicle in a vertical position.

The MST provides access to the spacecraft and launch vehicle and houses equipment to support spacecraft and launch vehicle prelaunch processing and launch. It includes service platforms, service fixtures, two freight/passenger elevators, and two service cranes.

Fig. 11.4. Proton rocket on launchpad at the Baikonur Cosmodrome. (Courtesy Lockheed Martin Space Systems)

Facilities

Baikonur Proton facilities include Building 92A-50, where Proton M, payload, and upper-stage processing occur; Building 92-1, where Proton K launch vehicle assembly and integration take place; and Areas 81 and 200.

The main buildings within the technical complex are the integration and testing facilities. Assembly and integration of the various stages of the Proton launch vehicle are carried out in the launch vehicle integration building. Spacecraft preparation, testing, and integration with the launch vehicle's fourth stage and the fairing are carried out either in Area 92 (Building 92A-50) or in Area 31 (Buildings 40/40D), both of which are spacecraft processing areas. Fueling and pneumatics pressurization are conducted in either Area 92 (Building 92A-50) or Area 31 (Building 44, the spacecraft fueling room). The launch vehicle's fourth stage and the mated spacecraft and fairing are transferred to the launch vehicle technical zone in Area 92 (Building 92-1), where they are horizontally mated to stages 1–3 of the assembled launch vehicle. The integrated launch vehicle is then transported to the Proton launch zone (Area 81) for erection, checkout, and launch.

The Proton launch vehicle processing room, in Building 92-1 at Area 92, is used to horizontally mate the assembled launch vehicle stages and strap-on elements. Building 92, approximately 50 m wide and 120 m long, includes the integration/assembly room and two laboratory annexes adjoining the assembly room. Building 92-1 contains heating, ventilation, fire-fighting systems, fire and security alarms, and special lighting. An overhead crane is available for handling the integrated launch vehicle.

Site 95 at Baikonur, sometimes referred to as "Proton City," contains living quarters for Proton workers. This area also includes three hotels for visitors and barracks for Russian military personnel.

Instrumentation

A new system has been installed that allows remote launch control and monitoring of spacecraft from one centralized location in Baikonur. This system allows customers to install their electronic test, command, control, and monitoring equipment in one location. The equipment needs no further transportation during different phases of the launch campaign. Overall campaign duration is thus minimized by elimination of transport, setup, and checkout time for equipment that would otherwise be required at different locations.

Fiber-optic cables that have been installed from the spacecraft control room at building 92A-50 lead to all major processing areas at the launch base, including the launchpad. This system provides direct optical fiber access, or can provide serial- or analog-to-optical and optical-to-serial/analog converters. In the system's fiber-optics video distribution system, video and communications data are also provided by fiber and routed to all processing areas.

Payload Processing

Commercial spacecraft are processed primarily at the Building 92A-50 processing facility. Tracks in room 102 of Building 92A-50 allow the railcars with the spacecraft containers and support equipment to be offloaded in a clean-room environment. From there the spacecraft containers are moved by crane and dolly into room 101, a class 100,000 facility. A state-of-the-art facility, Building 92A-50 is a comprehensive processing complex designed for simultaneous processing of spacecraft, upper stage, ascent unit, and Proton booster.

The spacecraft processing areas at the cosmodrome for Proton missions include Area 31, Buildings 40/40D; Area 31, Building 44; Area 81, launchpads 23 and 24; and Area 92, Buildings 92-1 and 92A-50.

Vehicle Integration

For Proton M/Breeze M launch vehicles, the integration of the encapsulated spacecraft and fourth stage to the launch vehicle is carried out in 92A-50, in Room 111. When launched on a Proton K, the encapsulated spacecraft and the fourth stage are transported from Building 92A-50 to Building 92-1 on an environmentally controlled railcar. The environmental railcar also provides for monitoring and recording of the spacecraft environments during transport, including acceleration loads. At Building 92-1, the spacecraft/upper-stage combination are integrated to the Proton's first three stages. When launched on a Proton M, this activity is conducted in Room 111 of Building 92A-50.

Soyuz

Launch Complex

Launchpad 5 at Launch Complex 1 and launchpad 6 at Launch Complex 31 are the two operational pads used for Soyuz launch vehicle preparation and launches at Baikonur. The launch table and customer bunker are the main facilities used on the launchpads. The Soyuz launch table includes all equipment, servicing towers, and access platforms necessary to support launch vehicle preparation and launch.

Facilities

Starsem has recently developed new Soyuz operations facilities at Baikonur, and most are located in Areas 31 and 112. Customers now have access to an 1158 sq m operations complex consisting of three separate areas. The complex is located inside the cosmodrome's Energia Hall, building MIK 112.

The hazardous processing facility (HPF) was designed for operations such as propellant filling and pressurization. HPF has a class 100,000 clean room, control room, and safety shower. The Upper Composite Integration Facility (UCIF), with a class 100,000 clean room, is where the spacecraft is integrated with the launcher's upper stage and fairing installation. UCIF has one control room for nonhazardous operations, an equipment airlock, and a personnel airlock. The payload processing facility contains two control rooms, a class 100,000 clean room dedicated to spacecraft processing, two personnel airlocks, and one equipment airlock.

Remote-Control Room

The remote-control room (RCR) is equipped with TV monitors from the HPF and UCIF video cameras. Intercom service is available, allowing direct communication with operators wearing fixed and mobile headsets. Power supply and environmental condition parameters from clean rooms can also be monitored by computer in RCR.

Hazardous-Storage Facility

The hazardous-storage facility is a dedicated facility designed for the storage of liquid propellant in tanks or drums, all of which are subjected to safety constraints. The surface area available is approximately 160 sq m for fuel and 150 sq m for oxidizer, and it includes fire- and leak-detection protection systems.

Zenit

Launch Complex

The Zenit launch complex at Baikonur includes spacecraft testing equipment, launch vehicle checkout facilities, technical support systems, spacecraft testing equipment, and ground technological equipment. The launch complex provides for installing the launch vehicle onto the launchpad, executing a complete testing cycle, fueling, and preparing the integrated launch vehicle prior to launch. The Zenit launch site at Area 45 contains two launchpads, but only one, Pad 45 Left, is currently operational. This pad was the site of the first Zenit launch in April 1985. Towers are located adjacent to the pad to protect it from lightning.

Zenit launch vehicle preparation is an automated process that provides reliability, performance, and safety with fewer personnel. An important feature is that no elements need be replaced after launch, which drastically reduces preparation time for the next launch.

Facilities

The Zenit technical complex is intended for receipt, storage, maintenance, and preparation of the launch vehicle; preparation of spacecraft in the clean room; launch vehicle and spacecraft mating; and checkout of the integrated launch vehicle and reloading to the transportation/installation unit for transfer to the launch complex prior to launching. The technical complex contains launch vehicle and spacecraft checkout facilities, technical-support systems, engineering facilities, and ground technological equipment. Spacecraft processing can be carried out in both Areas 31 and 254. In the Zenit assembly and test building, vacuum and antenna testing are conducted and checkout operations are completed. After Zenit stages arrive by rail from Ukraine, only about 76 hours are needed to assemble them prior to integration with the spacecraft. The integrated launcher is then transferred by rail to the launchpad.

Acronyms

HPF	hazardous processing facility
MST	mobile service tower
RCR	remote-control room
UCIF	upper composite integration facility

Points of Contact

Dnepr

Kosmotras
P.O. Box 7
Moscow, Russian Federation
123022
Voice: +7 (095) 745-72-58
Fax: +7 (095) 232-34-85
E-mail: info@kosmotras.ru

Proton

International Launch Services
1660 International Drive
Suite 800
McLean, Virginia 22102
U.S.A.
Voice: (571) 633-7400
Fax: (571) 633-7500

Soyuz

Starsem
5-7, rue François Truffaut
91042 Evry Cedex
France
Voice: +33 1- 698 70110
Fax: +33 1-607- 83199

Rockot

Eurorockot Launch Services GmbH
P.O. Box 28 61 46
Airport Center
Flughafenallee 26
D-28199 Bremen
Germany
Voice: +49 421 539-65 01
Fax: +49 421 539-65 00
E-mail: eurockot@astrium-space.com

Zenit

Yangel Yuzhnoye State Design Office
49008 Dnipropetrovsk
Krivorizka str., 3
Ukraine
Voice: 38 (0562) 42 00 22
Fax: 38 (056) 770 01 25, 38 (0562) 92 50 41
Teletype: 143294 deviz
E-mail: info@yuzhnoye.com

Tsyklon

United Start Corporation
2995 Airway Avenue
Costa Mesa, California 92626
U.S.A.
Voice: (714) 755-7427

Fax: (714) 545-7676
E-mail: info@unitedstart.com

References

"Baikonur Cosmodrome," *Design Bureau of Transport Machinery (KBTM)*, 2001, <http://www.kbtm.ru/english/complexes/baikonur.htm> (29 April 2005).

"Baikonur Cosmodrome," *ILS: International Launch Services*, last modification 28 January 2003, <http://www.ilslaunch.com/launchsites/Baikonur/> (29 April 2005).

"Baikonur Cosmodrome (NIIP-5/GIK-5)," *Russian Space Web*, <http://www.russianspaceweb.com/baikonur.html> (29 April 2005).

S. J. Isakowitz, J. P. Hopkins Jr., and J. B. Hopkins, *International Reference Guide to Space Launch Systems*, 4th ed. (AIAA, Reston, VA, 2005).

C. Lindborg, "Baikonur Cosmodrome Launch Facilities—Russia and Space Transportation Systems," *Federation of American Scientists, Space Project Policy, World Space Guide*, <http://www.fas.org/spp/guide/russia/facility/baikonur.htm> (29 April 2005).

Starsem, the Soyuz Company, <http://www.starsem.com> (29 April 2005).

M. Wade, ed., "Baikonur," *Encyclopedia Astronautica*, last modification 30 March 2005, <http://www.astronautix.com/sites/baikonur.htm> (29 April 2005).

12 Sweden

Esrange
• suborbital • 67.88 deg north • 21.1 deg east

Overview

Soon after its creation in 1962, the European Space Research Organization (ESRO) initiated a research program that was to use sounding rockets to study the atmosphere, the ionosphere, and the auroral phenomena. ESRO began a search for a high-latitude launch site for its sounding rockets, one preferably located north of the Arctic Circle. The site search eventually focused on an area in far northern Sweden, 43 km from the town of Kiruna. The large area, mostly unpopulated, would facilitate recovery operations, and the proximity of Kiruna would give site employees access to a well-developed town. Thus ESRO reached an agreement with Sweden in 1964 to establish the Esrange–ESRO sounding rocket launching range. Construction began in 1965, and launch operations began in November 1966 with the successful launch of a French-manufactured Centaure 1.

More than 150 sounding rockets were launched at Esrange from 1966 until June 1972, primarily by ESRO. Individual ESRO member states also sometimes conducted launches as part of national space programs. With the creation of the European Space Agency (ESA) in 1972, Sweden assumed formal control of Esrange, and since then the site has been owned and managed by the Swedish Space Corporation (SSC). Seven ESRO members made an agreement—known as the Esrange Special Project (ESP)—to contribute annually to Esrange in exchange for use of the site at a reduced charge.

Balloon launches were added to the activities at Esrange in 1974, and by the 1990s they accounted for the largest increase in site operations. Much of the increase in balloon activity occurred because balloons have proved especially useful in studying ozone depletion. Rocket and balloon activities are conducted and financed by the Esrange Andøya Special Project (EASP) within ESA. (EASP/ESA member states are France, Germany, Switzerland, Norway, and Sweden.)

The late 1970s saw a considerable increase in the number of sounding rocket launches. More than 300 sounding rockets have been launched at Esrange since Sweden assumed control in 1972. In addition to Centaure, sounding rockets launched from Esrange include the Nike-Apache, Nike-Cajun, and members of the Skylark and Black Brant series.

Esrange expanded its operations to include satellite operations in 1978, primarily as a result of ESA's desire for a satellite tracking station in a northern location. The Esrange station has proved especially useful in tracking satellites in polar orbits. Equipment for satellite operations has gradually expanded to include equipment for remote-sensing satellites, communications satellites, and scientific satellites. The SSC is considering expansion of Esrange's services to include launches to place satellites into polar orbits.

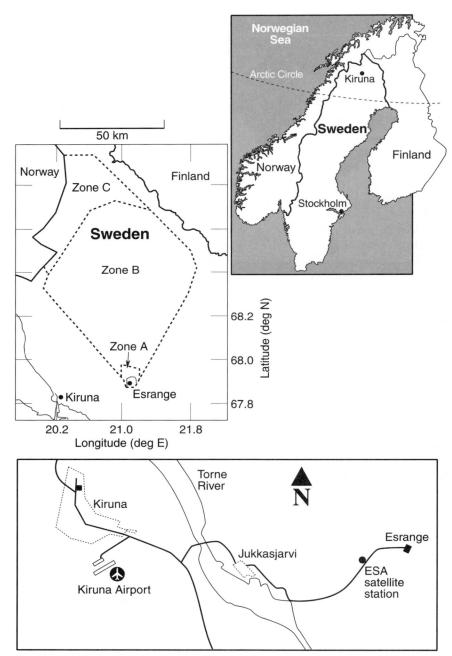

Fig. 12.1. Top right: location of Kiruna, Sweden; center left: rocket impact area north of the Esrange launch site; bottom: the route from Kiruna to Esrange.

Launch Site Description and Facilities

The Esrange Division (ED) is responsible for launch services and for operation of Esrange's technical facilities. Esrange has five permanent launchpads and three mobile pads, capable of launching large sounding rockets such as the *Skylark-12* and the *Black Brant IX*. The launching area for rockets includes a blockhouse, rocket and payload preparation halls, and chemical laboratories. An adjacent 100-m-high wind tower has wind speed gauges. The radar station, the satellite receiving station, and a GPS (Global Positioning System) reference station are atop a hill 2 km southwest of Esrange's main building. A ground observation station is on another hilltop about 1 km further west.

Close to the main building is the area for stratospheric balloon launchings, including two buildings for operations control and payload preparation. Balloons with a volume up to 2 million cu m can be launched from Esrange, which means that payloads weighing up to 2 metric tons can be carried to an altitude of 45 km. The balloon launchpad is 400 m long and 250 m wide at its widest point.

The Satellite Operations Division provides satellite support services, including TT&C (telemetry, tracking, and command) for launch, early orbit phase, and routine operations; data acquisition; data handling; engineering services; and operations and maintenance services.

Esrange's scientific instrumentation support facility provides support to all types of ground-based instruments used in sounding rocket and balloon scientific research experiments. Most of the instruments are stand-alone, but since 1996 Esrange customers also have had access to the Esrange Geophysical Information Services (EGIS) system, a multiple-instrument network that handles both live and archived geophysical data.

Impact Area

The rocket impact area is located due north of Esrange in the Swedish tundra region. This area is divided into three zones with a total area of 5600 sq km. Zone A, the impact area for boosters, can be extended when rockets with long-range boosters are launched. Zones B and C are impact areas for second and third stages as well as payloads, which are easily recoverable by helicopter, usually within an hour of impact. Zone C may not be used from May 1 to September 15. The nominal impact point normally chosen is 75 km north of the launchpads. The impact area for balloons covers the northern parts of Sweden, Norway, Finland, and Russia.

Central Complex

The four-story main building has a total floor area of 3930 sq m. The basement is used as storage for spare parts and also contains mechanical, electrical, and carpentry workshops. The ground floor houses offices for Esrange administration and technical facilities, a front desk, a switchboard, a kitchen, a canteen, two conference rooms (one holds 15 people; the other, 30), a library, and a lounge. The first floor has offices for operational staff, the operations center for sounding rockets, and rooms that contain timing, telemetry, and scientific instruments.

Associated Support Facilities

Arena Arctica, built in 1992, is a 5000-sq-m hangar at Kiruna Airport capable of housing a single plane the size of a Boeing 747 or several smaller planes. Its primary purpose is to support scientific measurements by aircraft. Arena Arctica also has offices, laboratories, a conference room, and a workshop. Offices and labs are equipped with modern communication equipment—telephone, fax, Internet access, and satellite channels. The hangar is fully equipped for ramp handling with towing trucks, stairs, high loader, air starter, airplane heaters, deicing equipment, and water service.

Acronyms

EASP	Esrange Andøya Special Project
ED	Esrange Division
EGIS	Esrange Geophysical Information Services
ESA	European Space Agency
ESP	Esrange Special Project
ESRO	European Space Research Organization
GPS	Global Positioning System
SSC	Swedish Space Corporation
TT&C	telemetry, tracking, and command

Points of Contact

Swedish National Space Board
Solna strandväg 86
Box 4006, 17104 Solna
Sweden
Voice: +46 8 6276480
Fax: +46 8 6275014
E-mail: spaceboard@snsb.se

Swedish Space Corporation, Esrange
P.O. Box 802
SE-981 28 Kiruna
Sweden
Voice: +46 980 720 00

References

"Esrange Launch Site," *Swedish Space Corporation*, <http://www.ssc.se/esrange/> (25 March 2005).

S. Grahn, *Sven's Space Place*, <http://www.svengrahn.pp.se/> (25 March 2005).

"Life in Space: Esrange and ESA," *European Space Agency*, last modification 20 October 2004, <http://www.esa.int/export/esaCP/ESA1GXLBAMC_Life_0.html> (25 March 2005).

J.-J. Serra, "ESRANGE (European Sounding Rocket Range)," *Rockets in Europe*, <http://www.univ-perp.fr/fuseurop/kirun_e.htm> (25 March 2005).

M. Wade, ed., "Kiruna," *Encyclopedia Astronautica*, last modification 16 April 2004, <http://www.astronautix.com/sites/kiruna.htm> (25 March 2005).

13 United States

Overview

Since the beginning of the space age in 1957, the United States has remained one of the two principal participants in space exploration, along with Russia. (Russia had been a part of the Soviet Union, until that country broke up in 1991.) Interest in rocketry by U.S. scientists and military can be traced back to the early 20th century. The United States was the site of the world's first rocket launch, a feat achieved near Clark, Massachusetts, on 16 March 1926 by Robert Goddard, the father of modern rocket propulsion. Goddard's early research provided the United States with the beginnings of what would ultimately develop into the world's largest space program.

The Soviet Union was the first country to launch an artificial satellite, *Sputnik 1*, in October 1957, and the United States was quick to follow with the launch of its own satellite, *Explorer 1*, on 31 January 1958. Although it initially appeared to lag behind the Soviets, particularly with a series of setbacks involving launch vehicle problems, the United States had been planning a presence in space for several years. In particular, proposals had already been made to launch a satellite in conjunction with observance of International Geophysical Year (July 1957–December 1958). However, the nature of the Cold War ideological struggle between the Soviet Union and the United States dictated a "space race" between the two countries that would be measured in terms of the firsts that each nation achieved, and the Soviets achieved a remarkable string of early space successes.

The nascent American space effort drew upon the expertise of rocket scientist Wernher von Braun and other German scientists and engineers, captured at the close of World War II, who had worked on the V-2 rocket program. American interest in rocket technology grew rapidly in the emerging atmosphere of the Cold War, particularly as awareness grew within the military and among leading scientists of the huge advances that the Germans had made in V-2 rocket development. Military leaders were anxious to develop rocketry because of its usefulness to weapons programs, while scientists were excited about the possibilities that rockets offered for exploration of the upper atmosphere and outer space. The National Advisory Committee for Aeronautics (NACA), founded in 1915, was responsible for much early American aeronautical and astronautical research. NACA was especially instrumental during the postwar years, in advancing knowledge of supersonic flight, which would be crucial to any future human spaceflight program.

Largely as a result of the public panic that followed early Soviet space successes, President Dwight D. Eisenhower authorized the formation of the National Aeronautics and Space Administration (NASA) as the successor to NACA on 1 October 1958. The new agency was charged with oversight of the civilian sector of the U.S. space program. Just six days later, on 7 October, NASA announced the beginning of Project Mercury, America's first crewed space program. Simultaneously, the Department of Defense (DOD) was organizing its own parallel space program. In 1958, the DOD assigned the Air Force the task of developing a crewed military space program that would utilize the piloted Dyna-Soar space plane, intended to be the world's first reusable spacecraft.

Fig. 13.1. Location of U.S. launch sites.

Following the presidential election of 1960, President John F. Kennedy sought a dramatic space goal that would pool the nation's skills and talents and overcome the perceived notion that the Soviet Union had become technologically superior to the United States. On 25 May 1961, Kennedy made the daring commitment to land and return an American astronaut from the moon before the end of the decade. This pledge was made just 20 days after the United States had launched its first Mercury astronaut, Alan Shepard, into space for a 15-minute suborbital flight and only six weeks after the Soviets had for the first time orbited a human (Yuri Gagarin) around Earth.

Despite public misgivings about spending huge sums to land men on the moon amid many domestic and international problems, Congress (and subsequent presidents Johnson and Nixon) eventually backed Kennedy's goal. In large part, this was because of the political perception that achieving a lunar landing before the Soviets did was an important Cold War objective. American space objectives received a major boost on 20 February 1962, when John Glenn became the first U.S. astronaut to orbit Earth.

Concurrent with the U.S. crewed space effort, American satellites, both civilian and military, made a remarkable series of successes:

- The early Explorer satellites discovered the Van Allen radiation belts.
- *Tiros 1*, the first weather satellite, was launched in 1960.
- *Telstar 1* became the first satellite to relay live television broadcasts between the United States and Europe, in 1962.
- *Surveyor 1* was the first U.S. craft to achieve a soft landing on the moon; *Lunar Orbiter 1* was the first U.S. craft to orbit the moon. Both events occurred in 1966.

- The Corona satellites began a remarkable program of worldwide photographic coverage for the DOD.
- The Defense Meteorological Satellite Program launched a series of military weather satellites that provided highly accurate weather forecasting in support of military planning operations.

The Air Force Dyna-Soar effort was cancelled by DOD in 1963, but hopes for a military crewed space effort continued with the approval of the Manned Orbiting Laboratory (MOL) program, also assigned to the Air Force, in 1965. The major purpose of MOL was the establishment of a crewed orbital laboratory to conduct surveillance and scientific experiments, but this program was also cancelled in 1969. The cancellation of MOL marked the end of attempts by the Air Force to have a military crewed space program operating simultaneously with NASA's civilian program.

After the conclusion of Project Mercury in May 1963, the Gemini program became the next major American space effort. Its goals were intended to demonstrate the principal spacecraft capabilities necessary to perform a lunar mission: for the two-man crew to rendezvous and dock with another vehicle and to complete a long-duration flight of up to two weeks, the time required for a flight to the moon and back. The first crewed Gemini flight took place in March 1965. By the conclusion of the program's 10 flights in November 1966, the United States had basically achieved parity with the Soviets in space performance capabilities. The final program of the moon landing effort was Apollo. It employed a three-person crew launched by a massive *Saturn V* rocket to achieve the goal of a lunar landing, which occurred on 20 July 1969. The final Apollo flight to the moon took place in December 1972.

Post-Apollo U.S. space efforts during the remainder of the 1970s included the launch of the first U.S. space station, *Skylab*, in 1973; the joint U.S.-Soviet Apollo-Soyuz Test Project in 1975, which provided a symbolic end to the space race and was the last crewed flight of the decade; and the initiation of the development of the space shuttle, the first reusable spacecraft, in an effort to make spaceflight more cost-effective. The overall shuttle program was officially designated the Space Transportation System (STS) by NASA.

The 1980s began with the promising launch of *Columbia*, the first space shuttle in the U.S. fleet, in April 1981. A series of successful shuttle missions followed, but the craft faced ongoing problems, including long delays in the refurbishment process. On 28 January 1986, the shuttle *Challenger* exploded shortly after liftoff, killing the seven crew members. The cause of the explosion was traced to faulty O-rings in the *Challenger*'s solid-fuel rocket boosters. In addition to highlighting the need for design changes in the rocket boosters, the subsequent investigation also pointed out the need for management reforms that would lead to increased safety procedures. During the course of the investigation the entire shuttle program was halted, and flights did not resume until September 1988, when the shuttle *Discovery* was successfully flown.

The U.S. space program experienced changing priorities during the 1990s. In addition to the ongoing shuttle program, the largest project of the decade involved participation in the construction of the International Space Station (ISS), the largest space project ever attempted. Smaller budgets also made it necessary to emphasize the robotic exploration of space, particularly planetary exploration. Some of the missions that did so included the NASA/DOD Clementine Project for mapping the lunar surface, the Mars Pathfinder and Sojourner Rover missions, and the launching of the Mars Polar Lander. In 1995, another space milestone was achieved when the Global Positioning System became fully operational, marking a major advancement in the history of navigation.

The space program encountered another setback in February 2003, when the shuttle *Columbia* broke apart over Texas as it was reentering the atmosphere near the conclusion of its mission. Once again, the nation's shuttle fleet was grounded while investigations were carried out to discover the cause of the accident. Shuttle flights resumed with the successful flight of *Discovery*, launched on 26 July 2005. *Discovery*'s crew delivered supplies to the ISS and tested new tools and procedures during the 14-day mission.

In January 2004, President George W. Bush announced his vision for NASA's role in the near future of space exploration. His proposals for the nation's space program include the return of astronauts to the moon no later than 2020, the establishment of a permanent lunar research base, the retirement of the shuttle fleet once construction of the ISS is completed, and the development of a new, crewed space vehicle to replace the shuttle. The return to the moon is seen as a possible stepping-stone to an eventual crewed flight to Mars. Bush called for spending $12 billion during the next five years on the proposed space effort.

The principal U.S. launch sites—the Western Range at Vandenberg Air Force Base in California, the Eastern Range and Kennedy Space Center in Florida, and the Wallops Flight Facility in Virginia—were all established by the federal government in the 1950s. They quickly became the major locations for space launches by the military and civilian agencies, and they remain so today. They are also the only U.S. facilities available for orbital launches, although Wallops is primarily used for suborbital missions. While the U.S. military and civilian agencies continue to be the primary customers of U.S. launch sites, the sites are also sometimes used by commercial customers.

In addition to these sites, the United States also utilizes the Poker Flat Research Range and the Kodiak Launch Complex, both located in Alaska, and the Reagan Test Site, also known as the Kwajalein Missile Range, located on Kwajalein Atoll in the Marshall Islands. Poker Flat, some 48 km northeast of Fairbanks, is operated by the University of Alaska's Geophysical Institute, under contract to NASA, and it has been the site for the launching of hundreds of suborbital scientific sounding rockets and more than 1500 meteorological missiles since its first launch in 1969.

The 688 ha (1699 acre) Kodiak Launch Complex, about 400 km south of Anchorage, is owned and operated by Alaska Aerospace Development Corporation.

Two launchpads support both orbital and suborbital launches, and Kodiak has been used by the U.S. Missile Defense Agency as a testing range for the nation's missile defense system. Construction of Kodiak was completed in 2000, and its first launch took place in September 2001.

The Reagan Test Site has also been integral to testing for the U.S. antiballistic missile system. Kwajalein has been used by the U.S. military since 1946, beginning with the testing of atomic bombs. Its remote location in the vast Pacific Ocean made it an ideal site for missile testing, which began in 1961. In addition, the atoll is near the impact area for target ICBMs launched from Vandenberg. More recently, the Reagan Test Site has been involved in the testing of interceptor missiles fired from Kodiak to test the new national missile defense shield.

Acronyms

DOD	Department of Defense
ISS	International Space Station
MOL	Manned Orbiting Laboratory
NACA	National Advisory Committee for Aeronautics
NASA	National Aeronautics and Space Administration
STS	Space Transportation System

Eastern Range
• orbital • 28.47 deg north • 80.53 deg west

Overview

The Eastern Range (ER), headquartered at Patrick Air Force Base (PAFB) in Florida, is operated by the U.S. Air Force's 45th Space Wing, which is responsible for assuring the nation's safe access to space. The ER oversight area for rocket launchings extends from the eastern coast of Florida all the way to the central Indian Ocean. Vehicle launch and operating facilities for ER are located at the Air Force's Cape Canaveral Air Force Station (CCAFS). These facilities are shared with the Kennedy Space Center (KSC), operated by the National Aeronautics and Space Administration (NASA). The sites are collocated on the central Atlantic coast of Florida, approximately midway between Miami and Jacksonville and about 34 km north of PAFB. Military and commercial launches occur at CCAFS, while space shuttle launches take place at KSC.

The name "Cape Canaveral" has been the source of much confusion throughout the years, because the two launch sites are in close proximity, and the name has sometimes been applied to both sites or used interchangeably with "Kennedy Space Center." The entire area was originally referred to as Cape Canaveral; then, following the death of President John F. Kennedy in November 1963, Air Force and NASA facilities were combined and renamed the John F. Kennedy Space Center, while the geographic area was renamed Cape Kennedy. In 1973, the name of the land mass was changed back to Cape Canaveral, and the Air Force facility was again known as Cape Canaveral Air Station (CCAS) while the NASA facility retained the name KSC. Subsequently the Air Force renamed the facility CCAFS in February 2000.

The origins of ER can be traced back to 1940, when the U.S. Navy established the Banana River Naval Air Station as a support base for antisubmarine sea patrols. The facility was transferred to the Air Force in 1948, renamed the Joint Long Range Proving Ground, and placed on standby status. Following reactivation in May 1950, the facility was again renamed, this time as Patrick Air Force Base; responsibility for developing and operating ER was assigned to the Air Force.

The first missile launch from the Cape took place on 24 July 1950 at what would later be known as Launch Complex (LC)-3. The missile was a German V-2 rocket called *Bumper 8*; it had an Army WAC-Corporal second stage. In October 1952, the Air Research and Development Command approved a plan to extend ER's seaward length to 8047 km to meet the requirements for missile testing. More launch complexes were constructed during the 1950s as U.S. missile programs expanded. After the U.S. space program began, following the launch of *Sputnik 1* by the Soviets in October 1957, CCAS became increasingly dedicated to space operations. The first U.S. orbiting satellite, *Explorer 1*, was launched from CCAS in January 1958.

When crewed spaceflights began in the early 1960s, NASA established its own launch operations center on Merritt Island. While launches continued from CCAS, it became apparent that NASA would need additional facilities for the larger rockets that were being planned. Recommendations were made in 1961 that development of a new NASA launch site be given high priority, and work proceeded quickly following President Kennedy's announcement that the U.S. intended to land a man on the moon before the end of the decade.

The vehicle used in the first crewed flight of Project Mercury, the *Freedom 7* capsule piloted by Alan Shepard, was launched from LC-5 on 5 May 1961. Project Mercury was the nation's first crewed space program. The second series of crewed flights was Project Gemini. The first crewed Gemini launch, using the *Titan II* launch vehicle, occurred in March 1965. All of the crewed Gemini flights took place during the period 1965–1966 and were launched from the Cape's LC-19. The Titan subsequently became the Air Force's workhorse launch vehicle at the Cape. The Air Force developed launch complexes 40 and 41 during the early 1960s to support its own crewed space programs, Dyna-Soar and the Manned Orbiting Laboratory (MOL), although both of these programs were ultimately canceled.

In the meantime, construction of the infrastructure for KSC's LC-39, the launch complex for the massive *Saturn V* booster used for the Apollo program, was completed by late 1966. Thirteen *Saturn V* launches took place from LC-39 between 1967 and 1973. Beginning in the mid-1970s, the two launchpads at LC-39 were modified to serve as the launch site for NASA's space shuttle. The launch of the first shuttle, *Columbia*, took place on 12 April 1981. The Air Force initiated efforts to stop its reliance on expendable launch vehicles (ELVs) and to begin

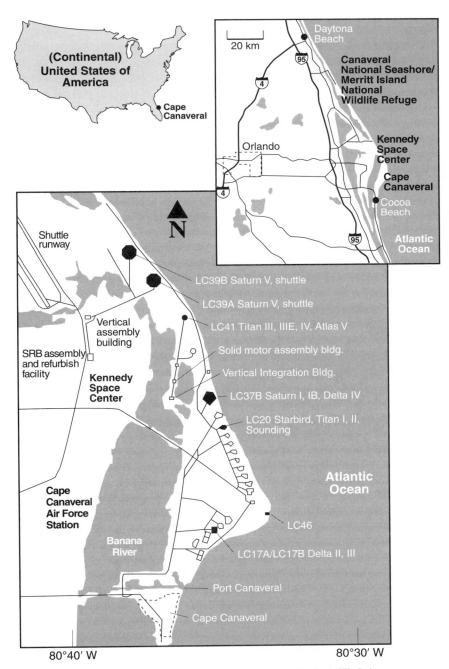

Fig. 13.2. The Eastern Range. (Site map © Mark Wade.)

integration of military and civilian space assets. All military spacecraft were now supposed to be launched aboard the STS.

The policy that all military spacecraft would be incorporated into the STS came to an end with the loss of the shuttle *Challenger* in January 1986, and the Air Force focused its efforts on establishing a mixed fleet of ELVs during the late 1980s and early 1990s. This period saw the addition of the *Atlas I*, *II*, and *III* and *Delta III* Medium Launch Vehicles family of expendable launch vehicles, and the transition of the Titan family of launch vehicles into the *Titan IV* launch vehicles.

The National Space Transportation Policy, announced in 1994, specifically assigned ELV development to the Air Force, while NASA was given responsibility for reusable launch vehicles. CCAFS was the site of the first *Atlas V* evolved expendable launch vehicle (EELV) launch, in August 2002, and the first *Delta IV* EELV later that year.

Launch Site Area Description

ER headquarters are located at PAFB, and the ER launch facility is located at CCAFS, about 111 km east of Orlando; it is situated on a barrier island between the Banana River and the Atlantic Ocean. The nearest major towns are Cocoa Beach and Cape Canaveral. The CCAFS location provides a natural safety buffer around the launch area, and it allows for launches over water into low-inclination and geosynchronous orbits. The range also includes instrumentation sites along the Florida coast as well as island stations and can support orbital as well as suborbital launches and mobile air and sea launches from the Atlantic Ocean area.

CCAFS encompasses 6390 ha (15,800 acres), although approximately 70 percent of its area has been preserved in a natural state. The barrier island parallels the mainland separating the Banana River and the Indian River Lagoon from the Atlantic Ocean. It is approximately 7.3 km wide at its widest point, and its elevation ranges from sea level to 6 m above mean sea level. CCAFS has 23 km of ocean coastline and 19 km of river shoreline.

Mission Capability

In addition to the local instrumentation and support at CCAFS, KSC, and PAFB, ER has resources at Argentia, Newfoundland (Canada); Jonathan Dickinson Park, Florida; Antigua Air Station, Antigua; and Ascension Auxiliary Air Field, Ascension. ER launches are restricted by the U.S. coastal land masses to the north and south, islands in the Caribbean, Bermuda, the northeast coast of South America, and the African continent. In general, vehicles must be launched in an easterly direction and on an azimuth that provides protection for land masses and populated areas from nominal spent-stage impacts, vehicle overflight, and any debris that might be generated as a result of destruct actions taken. Launch azimuths are restricted to a range from 37 to 112 deg. Missions flown here include equatorial and low-inclination orbital space launches, planetary launches, suborbital launches, and submarine-launched ballistic missile launches.

CCAFS has active launch complexes for *Delta II* and *IV*, *Atlas V*, and small ELVs. Also, ER supports submarine-launched ballistic missiles from designated locations in the North Atlantic. The Cape also has facilities for storing rocket motors; hazardous propellants; and liquid hydrogen, oxygen, and nitrogen; as well as facilities for assembling and testing most missile and payload components. The ER industrial area, a large service complex located on the center-west side of CCAFS adjacent to the Indian River, includes a dispensary, cafeteria, and fire station, and offices for military and contractor personnel supporting various launch efforts at the Cape.

Support Capabilities

Radar Systems

ER uses six classes of metric tracking C-band radars, comprising a total of 10 radar systems. ER also uses three NASA-operated radar tracking systems at Wallops Island, Virginia. The ER radar network provides real-time target position, aircraft vectoring, and trajectory and signature data.

All ER tracking radar systems are capable of beacon and echo tracking. Operational control and coordination of the radar resources is provided by the Single Point Acquisition and Radar Control (SPARC) system located at CCAFS. The SPARC system enables the range radar controller to control all 2400 bps acquisition data on the range, including some off-range systems, such as those at Wallops Island.

Besides controlling the designation data, the controller uses the SPARC system to monitor track and mode status of all ER radar systems. ER also uses S-band and X-band surveillance radar systems for range safety aircraft and ship control.

Optics

The ER optics capabilities include metric, engineering sequential, and documentary imagery. Metric optics systems provide two-dimensional position-versus-time data. These systems include intercept ground optical recorders, advanced transportable optical tracking systems, tracking cinetheodolites, fixed-site telescopes, and a laser imaging system. Engineering sequential imagery provides mission event-versus-time data such as umbilical disconnects, hold-down release, engine ignition, liftoff, and booster separation. These data are produced for range users. Documentary footage shot at CCAFS is used primarily for historical documentation, safety investigation, and public-information purposes. The footage provides a record of launch-related events, including transport, assembly, checkout, payload closeout, and the actual launch.

Telemetry Systems

Land-based telemetry facilities at the ER consist of two stations on the mainland, Tel-4 and the Jonathan Dickinson Missile Tracking Annex (JDMTA), and two downrange stations on the islands of Antigua and Ascension.

Tel-4

Tel-4 is located within the boundaries of KSC and is designated Tel-4/KSC, Station 19. This site is on the east side of the Banana River, west of the Cape Canaveral lighthouse. Tel-4 has been operational since 1966 and has capabilities for data acquisition, processing, and storage; preparation of computer-formatted magnetic tapes, tape copying, and playback; and provision of analog charts and recordings.

Jonathan Dickinson Missile Tracking Annex

JDMTA is located near Tequesta, Florida, about 161 km south of Cape Canaveral, in a portion of Jonathan Dickinson State Park. The site has been operational since 1987. It is equipped with 2.2–2.4 GHz antenna systems capable of tracking four separate targets. JDMTA has facilities capable of displaying, recording, and retransmitting data directly to Tel-4 for distribution to the Range Operations Control Center (ROCC) or to outside customer facilities.

Antigua

Downrange Station 91 is located on the island of Antigua in the British West Indies. The telemetry facility is located away from the main Air Force base at Antigua Air Station, adjacent to the southeast corner of the airstrip at Barnacle Point. The site, about 1250 nmi downrange from CCAFS, supports midrange launch trajectories. A new facility, designated the Central Instrumentation Facility, was recently constructed on a hill just north of the telemetry site.

Ascension

The most distant downrange station is Ascension, Station 12, about 5000 nmi downrange from the Cape on the island Ascension in the South Atlantic Ocean. The telemetry complex is on South Gannet Hill overlooking the east end of the airstrip runway. At about 222 m above sea level, the four antennas on the site have a field-of-view advantage into the distant offshore Sonobuoy Missile Impact Location System array. Real-time or near-real-time data is provided by satellite to CCAFS and Tel-4.

Communications

ER's extensive communications network consists of communication satellites, microwave links, high-frequency radio, and various landline links to connect the sites and stations of the range with each other and the world. The range also receives mission support communications services from, or provides mission support to, other agencies including NASA, the U.S. Navy, and the Air Force's 4950th Test Wing.

Operational control of ER communications is exercised by the communication control centers at each major station. These centers monitor, allocate, and maintain transmission quality of all on-base and off-base circuits and technical operations nets for each respective station. Antigua is the nodal point for the Caribbean area, while Ascension is the net control station for ship and aircraft operations in Africa and the Atlantic and Indian Ocean areas.

Fig. 13.3. Liftoff of STS-58 from KSC pad 39B on 18 October 1993. (Courtesy NASA.)

Range Weather Operations

The Range Weather Operations (RWO) unit launches balloons and weather rockets to gather vital atmospheric data related to launchings. RWO personnel also provide standard meteorological support for other Air Force units requiring their assistance. CCAFS instrumentation includes camera, command sites, optical sites and radar, and an antenna farm for UHF (ultrahigh frequency), VHF (very high frequency), and high frequency radio communications. Range communications transmitters are located at the Malabar Transmitter Annex in Palm Bay, Florida. The radar site at PAFB and the recording optical tracking instrument at Melbourne Beach (48 km south of the Cape) also provide instrumentation support for Cape launches.

Facilities

CCAFS contains a large number of facilities. It is divided into 11 areas, with the facilities connected to a coordinate system within each area. Following is a description of some of these facilities.

Explosive Safe Area (ESA)-60 (Building 54445)

This explosive safe facility was quantity-distance sited for NASA ELV hazardous payload operations, such as ordnance installation and propellant loading, and is located along the east shore of the Banana River. Its 2104 sq m of floor space houses two isolated bays, one on either side of a holding area. The facility is currently mothballed.

Dynamic Balance Lab

The 686 sq m Dynamic Balance Lab (Building 54446) is located just south of ESA-60 and is still used by NASA for spacecraft processing.

Range Communications Building (Building 1641)

This building houses communications links to all CCAFS facilities, instrumentation sites, KSC, and other off-range support systems. All ER communications are relayed through this location to the new ROCC by fiber-optic cable.

Satellite Assembly Building (Building 49904)

This 8148 sq m building is used for processing payloads for Delta, Atlas, and other vehicles.

Shuttle Solid Rocket Booster (SRB) Recovery Facility (Building 66250)

This facility acts as receiving point for SRBs recovered by the SRB retrieval vessels, the *Freedom Star* and *Liberty Star*, operated by Lockheed Martin Space Operations.

Facility 80700

This facility is a liquid-propellant disposal facility for contaminated hypergolic fuels.

Fuel Storage Areas (FSA) 1—5
The tanks in FSA-1, #77615–#77619, each hold 476 barrels of fuel. FSA-2 is a solids storage area with a 1019 sq m magazine for missiles, two 521 sq m magazines for motors, and 7 storage igloos from 107 sq m to 581 sq m for pyrotechnics, boosters, and other explosives. Adjacent to Space Launch Complex (SLC)-17, FSA-3 has 8 igloos, ranging in size from 90 sq m to 486 sq m, and a 392 sq m magazine for class A, B, and C ordnance storage. FSA-4 is a controlled parking area for transportable liquid storage tanks. NASA normally maintains a tank here capable of holding 61,425 L of liquid nitrogen. FSA-5, a solids storage area, has four magazines for class A, B, and C ordnance and missile storage.

Launch Complexes
Approximately 40 launch complexes have been built at CCAFS and KSC; most have been demolished or deactivated and are now used for other purposes or have simply been abandoned. The only active KSC launch complex is LC-39, used for space shuttle launches. All other pads are at CCAFS, which usually uses the term SLC for its complexes, and sometimes uses the older term LC.

Space Launch Complex 17
SLC-17 is a dual launchpad complex (SLC-17A and SLC-17B) built in 1956 for launching Thor missiles. The first Thor launch from Pad 17B took place on 25 January 1957, and Pad 17A supported its first Thor launch the following August. After the U.S. space program began in 1958, SLC-17 began supporting space launches. It was modified in the early 1960s to support a series of launch vehicles derived from the Thor booster. Thirty-five Delta launches took place at SLC-17 in a six-year period, from the beginning of 1960 until the end of 1965. The site currently serves as a launch complex for *Delta II* (Pads 17A and17B) and *Delta III* (Pad 17B only). Pad 17B was modified to support launches of the *Delta III* in 1997.

Space Launch Complex 20
SLC-20 originally supported *Titan I* and *Titan II* launches. It was later modified to support the *Titan IIIA* launch vehicle, and it supported four *Titan IIIA* launches prior to its deactivation in 1967. It was reactivated in 1987 and is currently operated by the Florida Space Authority for commercial space activities. The complex is now inactive and in pad closure status.

Space Launch Complex 36
SLC-36 is a dual-pad complex (SLC-36A and SLC-36B). It was originally constructed to support the Atlas/Centaur program and was operated by NASA until 1989. The site was built as a single-launchpad complex in February 1961, but a second pad (36B) was constructed during the period 1963–1964. Both pads served as launch sites for the *Atlas II*, while 36B also supported the *Atlas III*. The last launch from SLC-36, an *Atlas III*, took place in February 2005.

Space Launch Complex 37

SLC-37 was originally built to support *Saturn 1B* rocket launches, but was later mothballed. Beginning in 1995, The Boeing Company constructed a new launch site at SLC-37 to support launches of its Delta IV EELV. The first *Delta IV* launch from SLC-37 occurred on 20 November 2002.

Launch Complex 39

LC-39 is a dual-pad complex (LC-39A and LC-39B) originally constructed for *Saturn V*/Apollo launches. Currently the only active launch site at KSC, LC-39 is used for launching NASA's Space Transportation System, the space shuttle.

Space Launch Complexes 40 and 41

Both SLC-40 and SLC-41 were built in the early 1960s as part of an Integrate-Transfer-Launch (ITL) facility. They were originally designed to support *Titan IIIC* space missions, which began launches in June 1965 as part of the Air Force MOL program. Principal facilities in the ITL area include the solid motor assembly building, the solid motor assembly and readiness facility (SMARF), and the vertical integration building (VIB). Launch vehicles are assembled in the VIB and SMARF, shared by both launch complexes, and are transported by rail to the launchpad.

From the 1960s through 1990s, the two complexes supported many *Titan IIIC*, *Titan 34D*, and *Titan IV* launches. SLC-41 was deactivated at the end of 1977, but it was later upgraded and modified for the *Titan IV* program during the period 1986–1988. Complex 41 supported its first *Titan IV* launch in June 1989. Between 1990 and 1993, SLC-40 was also upgraded and almost completely rebuilt to support the *Titan IV/Centaur*, first launched in February 1994. However, SLC-40 is now in pad closure status. The last *Titan IVB* launch from SLC-40 occurred in April 2005.

SLC-41 was deactivated during the late 1990s. Following the construction of new launch facilities and additional modifications, which were completed in 2001, SLC-41 currently serves as the launch site for the *Atlas V* EELV.

Space Launch Complex 46

In 1985, SLC-43 was decommissioned, and SLC-46 was built on the site as part of the U.S. Navy's *Trident II* ballistic missile effort. The new launch complex, ordnance-certified in November 1986, supported the Cape's first *Trident II* test missile launch on 15 January 1987. After Trident launch operations moved out to sea in 1989, the complex was placed on standby status. Spaceport Florida Authority (now Florida Space Authority) received grants from the Air Force to modify SLC-46 to handle small commercial space launch operations for small solid rockets. SLC-46, which was opened for commercial space launches in May 1997, recently supported *Athena I* and *II* vehicles. Its last launch, an *Athena I* , took place in January 1999.

Associated Support Facilities

Argentia, Newfoundland

Argentia is located on the southeastern coast of the Canadian province of Newfoundland. ER has mobile C-band radar and command systems located on the grounds

Fig. 13.4. Aerial view of Launch Complex 39. At center is the vehicle assembly building. (Courtesy NASA)

of the decommissioned U.S. Naval Facility Argentia to support high-inclination launches. Leased landline circuits provide communications for operations.

Jonathan Dickinson Missile Tracking Annex

JDMTA was established in 1987 to replace the upper midrange resources that were lost when the Grand Turk and Grand Bahama Island facilities were decommissioned. It is located near Tequesta, Florida, about 161 km south of Cape Canaveral, in a portion of Jonathan Dickinson State Park. The site provides telemetry, command, radar, and communications from an integrated control facility. JDMTA can simultaneously track two separate vehicles. Communications with CCAFS are provided by wideband microwave and landlines.

Antigua Air Station

Antigua Air Station is located approximately 2011 km southeast of PAFB in the northern Leeward Islands of the West Indies. The 174 sq km island is the site of both the Air Station and a U.S. Navy facility. The Air Station provides telemetry, command, radar, and communications for both ballistic and space launch operations. Communication is provided by a cable system that extends from Antigua to Puerto Rico and the Virgin Islands, while satellite links provide additional communication to the mainland.

Ascension Auxiliary Air Field

Ascension Auxiliary Air Field, operated by the U.S. Navy, is the most southerly of the range facilities. It is located on the island Ascension, approximately 8047 km southeast of Cape Canaveral, in the South Atlantic Ocean. The airfield was originally built to support air operations during World War II, and later it supported the Snark and the Navaho weapon systems testing programs. Ascension continues to support Navy ballistic missile testing and the upper-stage tracking and telemetry data requirements for some orbital launches. Data and voice communications are relayed by satellite and high-frequency radio. The U.S. Air Force provides weekly air service between Ascension and PAFB.

Acronyms

CCAFS	Cape Canaveral Air Force Station
EELV	evolved expendable launch vehicle
ELV	expendable launch vehicle
ER	Eastern Range
ESA	Explosive Safe Area
ITL	Integrate-Transfer-Launch
JDMTA	Jonathan Dickinson Missile Tracking Annex
KSC	Kennedy Space Center
LC	Launch Complex
PAFB	Patrick Air Force Base
ROCC	Range Operations Control Center
RWO	Range Weather Operations
SLC	Space Launch Complex
SMARF	solid motor assembly and readiness facility
SPARC	Single Point Acquisition and Radar Control
SRB	solid rocket booster
STS	Space Transportation System
VIB	vertical integration building

Point of Contact

45th Space Wing/XP
Patrick Air Force Base, Florida 32925
U.S.A.
Voice: (321) 494-8869
E-mail: spacebiz@patrick.af.mil

References

"Cape Canaveral Air Station," *FAS Space Policy Project–World Space Guide*, last modification 16 June 2000, <http://www.fas.org/spp/military/facility/ccas.htm> (13 June 2005).

Research Triangle Institute, Center for Aerospace Technology (CAST), Florida Office, 45th Space Wing/Patrick Air Force Base: Launch Site Safety Assessment, Report No.: RTI/08087.002/TASK 1.2-06.0F (Federal Aviation Administration,

Washington, D.C., 2002), <http://ast.faa.gov/files/pdf/ER_Final_lssa_06_08_02.pdf> (13 June 2005).

S. J. Isakowitz, J. P. Hopkins Jr., and J. B. Hopkins, *International Reference Guide to Space Launch Systems* (AIAA, Reston, VA, 1999).

Patrick Air Force Base Web site, <https://www.patrick.af.mil/>.

R. Svirkas, ed., "Cape Canaveral Air Force Station Virtual Tour," last modification 5 April 2005, <http://www.robsv.com/cape/> (13 June 2005).

D. Day, "Spaceflight: Cape Canaveral," *U.S. Centennial of Flight Commission*, <http://www.centennialofflight.gov/essay/SPACEFLIGHT/KSC/SP46.htm> (13 June 2005).

J. Tomei and J. Gin, "Launch Vehicle Overview" class, Sections 12 and 13, The Aerospace Institute of The Aerospace Corporation, El Segundo, CA.

J. F. Wambolt and J. F. Kephart, "A Complete Range of Launch Activities," *Crosslink* **4** (1) (Winter 2002/2003).

Western Range
• orbital • 34.67 deg north • 120.62 deg west

Overview
The Western Range (WR) is located at Vandenberg Air Force Base (VAFB), situated in Santa Barbara County on the central coast of California, about midway between San Diego and San Francisco. VAFB is one of the two principal U.S. space launch sites.

WR, originally known as the Western Test Range, is the headquarters of the Air Force's 30th Space Wing (30 SW), which oversees all launch activities on the West Coast. Vandenberg is the nation's third largest Air Force base, comprising some 40,000 ha (98,840 acres).

In March 1941, just before the United States entered World War II, the Army acquired approximately 34,803 ha (85,963 acres) of this land to establish a new training base. Most of the property was purchased outright, while smaller parcels were obtained by license, lease, or easements.

Actual construction of Army facilities began in September 1941. The installation was quickly activated on 5 October and named Camp Cooke in honor of Maj. Gen. Philip St. George Cooke. Camp Cooke was deactivated after World War II, but it was opened again in 1951 during the Korean War (1950–1953).

In November 1956, the secretary of defense ordered the Army to turn over North Camp Cooke to the Air Force to establish a missile-testing and launch facility. (Subsequently the balance of Camp Cooke acreage was turned over to the

Navy for use as the Point Arguello Launch Facility, which was in turn transferred to the Air Force.) In May 1957, the Air Force initiated construction of the base infrastructure and launchpads, beginning with the construction of seven Thor intermediate range ballistic missile (IRBM) launchpads, including Space Launch Complex (SLC)-2. On 4 October 1958, Cooke AFB was renamed Vandenberg AFB in honor of the late Gen. Hoyt S. Vandenberg, the Air Force's second chief of staff. Space launch operations at VAFB began with the launch of a Thor IRBM on 16 December 1958.

On 28 February 1959, the world's first polar-orbiting satellite, *Discoverer I*, was launched into space aboard a Thor/Agena booster combination. By 1960, Thor IRBM flights and Thor/Agena orbital missions were well under way at WR. By June 1960, WR had flight-tested 45 ballistic, orbital, and probe launch vehicle systems. Titan intercontinental ballistic missile (ICBM) test flights began with the first *Titan I* demonstration flight in early 1961. WR also provided support to several emerging programs such as Nike-Zeus Target and Satellite and the Missile Observation System. An extensive Atlas flight-test period in support of the Advanced Ballistic Reentry Systems program ultimately involved 81 launches during a 10 yr period.

The final major land acquisition at Vandenberg occurred on 1 March 1966, after the Air Force had announced plans to construct Space Launch Complex 6 (SLC-6) on former Sudden Ranch property for the Manned Orbiting Laboratory (MOL) program, which had been approved the previous year. Located south of the former naval missile facility at Point Arguello, the property consisted of nearly 6070 ha (15,000 acres). Following construction of the MOL launch complex at SLC-6, the Air Force proposed a plan to launch seven crewed MOL missions from VAFB on a modified *Titan IIIM*.

Four uncrewed launches and the first crewed one were planned for December 1969 from WR. However, as a result of program technical problems, advances in space systems, cost overruns, and changing national priorities, the MOL program was canceled in June 1969, leaving the nearly completed SLC-6 unused. Despite the cancellation, launch vehicle activity at WR continued at a busy pace during the 1970s, with some 30–40 launches per year using Thor/Agena, Thorad/Agena D, Thor, *Titan IIIB*, *Atlas F*, and Scout, and Delta.

In 1979, modification work began on SLC-6 in preparation for Air Force shuttle launches from WR. The Air Force relocated the Shuttle Activation Task Force from Los Angeles to VAFB that year. In early 1985, after major construction had been completed and systems installed, the space shuttle flight verification vehicle Enterprise was brought to VAFB and erected to perform form, fit, and function tests. After an expenditure of about $4 billion, SLC-6 achieved initial launch capability status in October 1985, and the first West Coast shuttle launch was scheduled for 1986. However, the disastrous explosion of the shuttle *Challenger* in January 1986 resulted in the cancellation of all shuttle-related operations at WR.

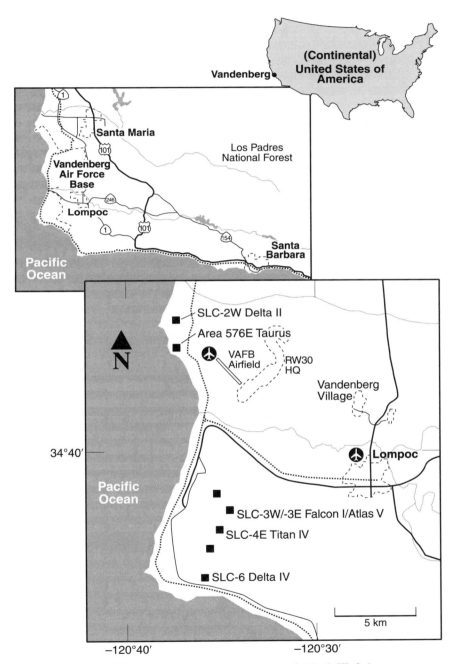

Fig. 13.5. The Western Range. (Site map © Mark Wade.)

Following the loss of *Challenger*, the Air Force needed nonshuttle options for access to space. It reactivated expendable launch vehicle programs and acquired additional Titan launch vehicles from Martin Marietta. The National Aeronautics and Space Administration (NASA) and the Air Force jointly decided to consolidate shuttle operations on the United States' Eastern Range at Cape Canaveral, Florida, and the SLC-6 facilities were again slated to be unused. In 1994, Lockheed proposed using SLC-6 to test and launch its newly developed small launch vehicle, the Lockheed Launch Vehicle. The Air Force agreed and a lease was signed, giving Lockheed access to SLC-6. The first launch attempt, in August 1995, failed to place its payload in orbit but did bring first "fire" to SLC-6. After the merger of Lockheed and Martin Marietta, the vehicle was renamed the Lockheed Martin Launch Vehicle (LMLV) and still later called Athena.

Lockheed Martin subsequently moved Athena operations to the Kodiak Launch Complex on Kodiak Island, some 400 km south of Anchorage, Alaska, and SLC-6 was leased to The Boeing Company for the Delta IV Evolved Expendable Launch Vehicle program. In 2000, Boeing began modifications on SLC-6 to transform it into the West Coast home for the *Delta IV* family of launch vehicles. In addition to modifying the launch site, Boeing constructed a new horizontal integration facility and built a new launch control center within the already existing Remote Launch Control Center (RLCC) building located on North VAFB.

During more than four decades of operations, WR has supported more than 1023 space launches, including 174 Thor, 224 Atlas, and 120 Titan, four Scout, six Pegasus, and five Taurus.

Launch Site Description

VAFB is located on a prominent headland extending into the Pacific Ocean near Point Conception. The east-west orientation of portions of the headland allows for launchings to the south and southwest for polar and other high-inclination orbits without land overflights. The coastline of VAFB extends about 56 km with both west- and south-facing beaches. Nearby communities include Lompoc and Santa Maria. The topography of VAFB encompasses rugged coastal bluffs, dunes, beaches, and mountainous terrain. The long coastline and remote location of WR are the major reasons for its development into a major launch facility.

The West Coast Offshore Operating Area (WCOOA) is an aeronautical, ballistic-missile, and guided-missile test area along the West Coast. The 30 SW manages it as an adjunct to WR. Most testing is conducted off the California coast near VAFB, although the extensive test range, approximately 322 km wide and more than 1609 km long, extends from the Mexican border north to the Gulf of Alaska. WCOOA is a combination of restricted and warning areas and Federal Aviation Administration–monitored airspace.

The mission of the 30 SW is to conduct and support space and missile launches, operate WR, respond to worldwide contingencies, and serve as the host unit to the Vandenberg community. The 30 SW, through WR, provides real-time command and control of air-, sea-, and ground-based assets, ensuring the timely and safe conduct of spacelift operations in support of Department of Defense (DOD), NASA, and commercial space requirements.

WR is one of 20 DOD facilities included in the Major Range and Test Facility Base, and it is the only range that launches operational ICBMs. WR also supports aeronautical flight testing. It is a composite of all the resources and assets necessary to support space and missile launches and other test activities within an operational area ranging from the California coast to the Indian Ocean.

Launch Complexes

VAFB has five active, fixed launch complexes from which space launches are conducted. These are SLC-2 (*Delta II*), SLC-3E and W (*Atlas V* and *Falcon I*), SLC-4E (*Titan IV*), SLC-6 (*Delta IV*), and Area 576E (Taurus). In addition, the Pegasus launches utilize the base's airfield and 4572 m runway and are performed from the offshore aeronautical airspace. With the exception of SLC-2W and Area 576E, all space launch facilities are located on the southern portion of the base. SLC-6 has been modified to accommodate the Boeing *Delta IV* launch system and was declared operational in April 2005.

SLC-2 (Delta II)

SLC-2, located along the coast on the northern portion of VAFB, is the West Coast launch site for the *Delta II* expendable launch vehicle. SLC-2 was built for the Air Force in 1957 for its IRBM testing program. The complex launched the Thor IRBM in various Air Force testing programs during its early years. NASA acquired SLC-2E in 1962 and SLC-2W in 1969. Minor modifications were made to both pads so that NASA could launch its Delta rocket. As the need to launch polar-orbiting satellites lessened, NASA abandoned SLC-2E in 1975 and concentrated its operations at SLC-2W. In the same year, SLC-2E reverted to control of the Air Force and was decommissioned and stripped of all its equipment and salvageable materials. SLC-2 was refurbished in 1992 to accommodate the larger *Delta II* model. Among the more recent upgrades to the complex is an automated advanced launch control system, which provides greater availability of system status information to all operators through a commercial workstation on a high-speed fiber-optic computer network.

SLC-2 consists of launchpad (SLC-2W), blockhouse, launch operations building, horizontal processing facility, and other facilities necessary to prepare, service, and launch the Delta vehicle. The *Delta II* launch vehicle is fully integrated on the launchpad. The mobile service gantry and fixed umbilical tower were modified to accommodate the new, taller *Delta II* launch vehicle and larger strap-on solid rocket motors. The upper-level enclosure, or "white room" area, provides protection for the spacecraft, checkout equipment, and personnel while on the launchpad. Launch operations are controlled from the Remote Launch Control Center (RLCC), which is equipped with vehicle and ground systems monitoring and control equipment.

The horizontal processing facility, located near SLC-2W, is used for processing the Delta first and second stages. In the facility, which is south of the blockhouse, prelaunch processing is performed, as well as preparations for erection and transport to the launch mount.

A solid motor building is located near the SLC-2 area for processing of the strap-on solid rocket motors. The facility has been modified to handle both the Castor IV A and the larger graphite-epoxy motors. It is used for receiving, inspection, buildup, and preparation for erection and transportation to SLC-2W.

SLC-3E (Atlas V) and SLC-3W (Falcon I)

SLC-3 is located on the southern portion of VAFB. The launch complex initially consisted of two launchpads (east and west), commonly referred to as SLC-3E and SLC-3W, which were located about 533 m apart. The pads were built (under a Navy contract) for the Air Force in the early 1960s, to launch *Atlas D/Agena*, and subsequently modified to launch Thor, *Atlas E/F/H*, and the *Atlas II* family of boosters. The first *Atlas E* launch from this facility occurred in July 1961.

The decision in 1992 to modify SLC-3E from an *Atlas I* complex to one for the Atlas II family of launch vehicles derived from a perceived need to establish a medium launch vehicle capability at WR. The Air Force and the National Reconnaissance Office initiated development of the *Atlas/Centaur* capability at VAFB as a major investment keyed to using the medium booster for a new class of payload. The broad intent was to provide medium-lift capability to launch-surveillance, communications, and exploratory satellites that support commercial, DOD, and other national requirements.

Initial operating capability for the new facility was achieved in September 1997, and the first *Atlas IIAS* (AC-141) launch from SLC-3E took place in December 1997, with the launch of the $1.3 billion NASA Earth Observing System–Terra satellite. Beginning in 2003, SLC-3E was again modified for use by the *Atlas V* EELV, with a first launch scheduled for late 2005.

SLC-3W also consists of a single pad. It was originally built by the Air Force in 1960 for *Atlas/Agena* missions. Although it was decommissioned in the late 1990s after the last *Atlas E* launch, it has been reactivated and is currently leased from the Air Force by the SpaceX corporation for launches of *Falcon I* vehicles. The major facilities of SLC-3W include an umbilical tower, a support building, a mobile service tower, and propellant and gas storage areas.

SLC-4E (Titan IV) and SLC-4W (deactivated)

SLC-4E and -4W were part of a major facility overhaul on VAFB. To maintain the required West Coast launch capability of classified payloads, the direction was given to upgrade the Titan launch vehicle to accommodate the larger and heavier shuttle-class payloads that had been scheduled to fly on the shuttle. Modifications to the *Titan III* launchpad at SLC-4E were determined to be the most cost-effective and schedule-supportive approach to regain access to space. SLC-4E returned to operations in March 1991 with the first launch of the *Titan IVA* vehicle. The first *Titan IVB* launch took place in May 1999. The last such vehicle is on the pad, with a launch date projected for October 2005.

SLC-4W was the *Titan II* launch facility, located on the west end of the SLC-4 complex near the coastline. SLC-4W began operations with the *Titan II* in the mid-1980s and supported the launch of a variety of payloads, including Landsat, Defense Meteorological Satellite Program (DMSP), and National Oceanic and Atmospheric Administration satellites. The last *Titan II* launch occurred in October 2003, and SLC-4W is now inactive.

SLC-6

SLC-6 is located on the southern portion of VAFB near Point Arguello. Construction began in March 1966 to support launches of the Air Force's MOL. The MOL program was canceled in June 1969, and SLC-6 remained unused for nearly 10 years. In January 1979, a six-year reactivation effort began to transform SLC-6 into a West Coast space shuttle launch facility, but West Coast shuttle launches were canceled following the loss of *Challenger* in 1986. Once again, SLC-6 was temporarily abandoned.

In 1990, SLC-6 was designated the launch site for a *Titan IV* vehicle with a Centaur upper stage. It was quickly dubbed the Titan Centaur Launch Complex (TCLC). Development of TCLC progressed through concept evaluation and preliminary design before it too was canceled.

Lockheed Martin used SLC-6 to launch *Athena I* and *II*. From 1995 to 1999, four launches took place.

In 2000, Boeing began to transform SLC-6 into the West Coast home for the *Delta IV* family of launch vehicles. In addition to modifications to the launch site proper, a new horizontal integration facility was constructed and a new launch control center was added to the existing RLCC building located on North VAFB.

Area 576E (Taurus)

Area 576E, formerly a launch site for ICBMs, is relatively austere, with few permanent structures. Located south of SLC-2, it contains a launchpad surrounded by security fencing, lighting, and camera towers, and it is supported by facility power and the VAFB communications network. Area 576E is currently used by Orbital Sciences Corporation for the Taurus launch vehicle.

Remote Launch Control Center

Until 1986, countdown and launch were controlled from local blockhouses. This had been the case since the first rockets were launched from VAFB. However, the explosion of a *Titan 34D* shortly after liftoff from SLC-4E forced a change. With the addition of solid rockets to launch vehicles, risks had grown but had been, for the most part, underestimated or ignored. When the large solids on the *Titan 34D* exploded, sizable fragments caused damage to SLC-4 and started numerous fires. One big fragment, which could have penetrated the blockhouse, landed very close to it, and personnel were confined there for a number of hours until danger had passed.

The Aerospace Corporation subsequently conducted a study for the Air Force and concluded that use of a centralized location for all site countdowns and launches would be preferable to control by local blockhouses. Building 8510, which could house launch control facilities for a number of sites, was selected for the location of the RLCC.

Booster Processing and Storage

The 30 SW has several booster-processing facilities to support government, military, and commercial programs. Military Minuteman II and III and Peacekeeper programs have their own dedicated booster-processing facilities, as do *Titan II* and *IV*, *Delta II* and *IV*, and *Atlas V*. Commercial booster-processing facilities are provided on an excess-capacity basis. Taurus commercial booster processing is done in the missile assembly building; Pegasus booster processing, in Building 1555. Lockheed Martin does LMLV processing in the SLC-6 shuttle assembly building. Commercial Delta launch vehicles are processed in Building 1670.

Instrumentation

WR instrumentation sites are located along the Pacific coast at VAFB, Pillar Point Air Force Station, Anderson Peak, and Santa Ynez Peak. Midrange instrumentation is in Hawaii.

Communications

The 30th Communications Squadron provides base-level communications and computer support, wide area network information transfer, training, equipment maintenance, and direct customer service. It maintains and operates DOD computers, radios, radars, and meteorological and navigational systems, as well as communications centers located both on- and off-base, and it interfaces with the Federal Communications Commission. It also assesses the impact of integrating new communications computer systems into the basewide infrastructure. The unit provides small computer support, operates the base video teleconferencing center, provides photographic services, operates the base motion-picture laboratory, and operates high-speed motion-picture tracking for both space and ICBM launches from WR.

Data Collection and Analysis

Systems are available in launch control centers, in customer-provided facilities, and in the Western Range Control Center (WRCC) to allow customers to assess launch data during real time or after operations are completed. WRCC's launch operations control center has several consoles on which customers may view real-time video and telemetry status information. The "quick look" display area has dedicated rooms for customers to view real-time and postmission display of telemetry data. Postoperational data can be processed in standard operational data-item manual and best-estimate-of-trajectory formats.

Telemetry

WR provides telemetry coverage for space launches, ballistic launches, and aeronautical operations at two permanent sites, the Vandenberg Telemetry Receiving Site and Pillar Point. The telemetry systems can support customer frequencies in the L- and S-band regions, downlink video, pulse code modulation (PCM), and analog data. Each telemetry site can process PCM rates up to 5 Mbps. The received telemetry data are relayed in real time to the telemetry processing facility for display to range safety and customers. The data are also recorded on analog magnetic tape for postflight processing.

Optics
The 30 SW operates and maintains optical systems at WR to provide high-quality video and film records of launch operations. Three permanent optical sites are Tranquillion Peak, Santa Ynez Peak, and Anderson Peak.

Command Control
The Missile Flight Termination Ground System, in cooperation with a missile-borne receiver system, is used in support of in-flight missile flight control requirements for the 30 SW and the Naval Air Warfare Center. It comprises a central control processing system and five remote command control transmitter sites.

Airfield Operations
The VAFB airfield runway is 4572 m long (with a 305 m paved surface overrun on each end) and 61 m wide, and it is constructed of cement 38–43 cm thick. The runway is equipped on both ends with a CAT I instrument landing system and an approach lighting system. Other navigational aids include a tactical air navigation equipment system. An air surveillance radar is on the field. Four 23 m wide taxiways serve the aircraft parking apron located on the southeast side of the runway.

Weather Instrumentation
The 30th Weather Squadron supports all space and ballistic missile launches from VAFB and provides a 24 h meteorological watch that includes forecasts, observations, advisories, and warnings to enhance base resource protection. The 30th Weather Squadron also provides daily weather briefings to Headquarters 14th Air Force to support global operations in all military theaters. Worldwide weather information is received at the 30 SW from many sources, including geostationary and polar-orbiting satellite imagery. Weather instrumentation at VAFB includes wind towers, acoustic sounders, a WSR-88D Doppler weather radar, an ionospheric sounder, a Doppler windprofiler, an automated weather distribution system, and lightning detection and location systems.

Payload Accommodations
The 30 SW provides payload processing for both government and commercial payloads. Payload processing facilities are designed to support different payload processing requirements at VAFB. Available facilities include a class 10,000 clean room, hazardous fueling facilities, and spin test facilities. Four facilities have been approved to process hazardous hydrazine fuels and bipropellants. Two spin test facilities are available at VAFB for dynamic balancing that can accommodate up to a 2727 kg payload. Many facilities contain cranes for payload transport. Most payload processing facilities use a 5–10 ton crane; however, cranes as large as 75 tons are available.

Small Payload Processing Facility—Building 596
This 1244 sq m facility, constructed of wood and steel, was built in 1962 to support payload processing for the Navy's Scout launch vehicles. No special capabilities are provided other than two half-ton cranes for equipment handling. Category 1.4 pyrotechnic devices may be handled at this facility. Electrical test equipment for processing must be brought to the facility along with any other unique processing equipment. At present this facility is unused.

Missile Space Research Test Facility—Building 660
This facility, also known as the Naval Research Laboratory (NRL) facility, is owned by NRL and operated by its contractor, Allied Signal Corporation. The building, a 2280 sq m steel structure built in 1965, has a high bay upgrade that was added in 1976. Two high bays are available, one with a 5 ton crane and another with a 1 ton hoist. A portable 3 by 3.6 m clean room is available for special processing operations. Current operations include tracking of specified Navy satellites and payload processing. Backup power, provided by a 200 kW diesel generator, is available.

NASA Spacecraft Laboratory (or NASA Payload Facility)—Building 836
Building 836, a 16,416 sq m facility built in 1960, supports processing for NASA polar-orbiting spacecraft. Upgrades added laboratory and microwave transmission capabilities in 1965. The two large spacecraft laboratories with class 100,000 cleanliness ratings can accommodate spacecraft weighing up to 5 tons with diameters up to 3 m. D53-, S- and X-band RF, and video links, are available for data transfer to the range network control center in building 7011. Approximately 3022 sq m of administrative space is available. No hazardous material provisions exist. Some limitations are imposed on the cranes because of their age, as parts replacement is minimal or nonexistent. The building, a multipurpose facility, also houses the NASA telemetry, ground station, and engineering support area, and the Delta II launch vehicle staging area.

Spin Test Facility—Building 995
This facility is a 353 sq m block wall structure with steel siding whose primary function is spacecraft spin balance testing. Spacecraft weighing up to 727 kg can be handled on the Gisholt vertical spin balance machine. The facility contains a high bay with a 14 ton bridge crane and a compressor room. Telephone, two-channel video, and radio frequency data links are provided for communication to the NASA ground station in Building 960. Approximately 61 sq m of administrative space is provided. Small rocket-motor storage and pyrotechnic device handling are permitted at this facility. This compound transitioned from NASA to the Air Force following completion of the Scout program in 1994. It is currently used for reentry vehicle processing for the Peacekeeper ballistic program and has been equipped with a newer, although smaller, spin table.

DMSP Payload Test Facility—Building 1559

Building 1559, the DMSP assembly checkout area, is a 6691 sq m facility built in 1963 and modified to consolidate the modular payload test facility formerly located at building 1768 during 1991 and 1992. It supports DMSP (Defense Meteorological Satellite Program) processing and other programs such as Landsat, Peacekeeper, and Minuteman. Two class 100,000 clean rooms are available with crane capacity up to 5 tons. Hydrazine handling capabilities are available, and class 1.1 pyrotechnic handling is allowed. A 305 sq m anechoic chamber is located in the facility. Storage space for up to three DMSP spacecraft is also available.

NASA Spin Test Facility (or NASA Hazardous Processing Facility)—Building 1610

A variety of NASA spacecraft are processed in this 1204 sq m facility built in 1966. An environmental equipment room was added in 1990. One class 100,000 clean room rated high bay is available for processing spacecraft weighing up to 5 tons. Pyrotechnic and solid motor hazard ratings are applied to this facility. Hydrazine and oxidizer handling capabilities are available, including a spill trench located in the high bay adjacent to the spin table. Data communications are provided via S- and X-band microwave, VHF (very high frequency), video, and operational intercoms. A spin balance table is capable of handling payloads weighing up to 2727 kg. Both technical and facility grounding are available.

Associated Commercial Support Facilities

The 30 SW currently supports commercial launches from VAFB. Some of the companies with facilities at Vandenberg include Astrotech Space Operations, Orbital Sciences Corporation, and California Commercial Spaceport Incorporated. The following information contains brief descriptions of the commercial facilities at Vandenberg.

Astrotech Payload Processing Facility—Building 1032

This facility is the first to be fully developed, from the ground up, as a commercial payload processing facility on VAFB. Located on 24 ha of leased land, the facility was designed to provide space for satellite final assembly and checkout. It can handle both nonhazardous and hazardous operations including propellant loading, solid motor processing, and ordnance activities. It also provides the capability for payload stacking and fairing encapsulation. Currently providing more than 9754 sq m of floor space, the facility includes two clean room high bays, a clean room airlock, two control rooms, and two garment change rooms. Both bays and the airlock have a class 100,000 clean room rating. The facility provides three overhead cranes: 10 ton, 11 m hook height (HH); 15 ton, 15 m HH; and 30 ton, 17 m HH. The building incorporates an explosion-proof design and is EPA- and OSHA-compliant.

Pegasus Vehicle Assembly Building—Building 1555

This facility was built in 1965 and upgraded in 1972. Formerly used as a PPF, it was occupied in September 1993 by Orbital Sciences Corporation after extensive building modifications and became its Pegasus assembly building. As such it provides both vehicle and payload integration capabilities for the Pegasus. Two high bays and crew work areas comprise 9129 sq m. A portable 3 by 7 m class 100,000 clean room is available in the high bay area. Hazardous processing for hydrazine and pyrotechnic devices is available; however, all other processing must be halted during any hazardous operations. Up to four vehicles and a single payload can be processed simultaneously, but only one spacecraft can be fueled at any time. Current plans are to employ this facility for booster and payload integration only and to accomplish payload processing elsewhere.

Integrated Processing Facility—Building 375
California Commercial Spaceport Incorporated (CCSI), a for-profit California corporation for processing government and commercial payloads, operates Building 375, the Integrated Processing Facility, located at SLC-6. The facility is a 31,647 sq m reinforced concrete structure containing three payload cells, each with seven service levels and access from an erection room that provides lifting service with a 75 ton bridge crane. Cell dimensions are 11 by 13 by 21 m, and each cell has an access-door area of 7 by 22 m. Payload access to the facility is through a horizontal airlock that contains two 25 ton cranes.

The entire processing area has a clean room rating of class 100,000 with associated heating, ventilation, and air-conditioning (HVAC) filtering. The payload cells are each provided with a 5 ton bridge crane for local lifting services. Cell I is a 100 dB RF shielded room that provides an acoustic rating of STCA5.

Piping systems are installed to handle hydrazine propellants, and the facility has spill containment with a 94,635 L storage capability. Both facility and isolated single-point technical grounding are provided at each level in the service cells and erection room. Power systems include both one- and three-phase 120 V service and three-phase 480 V service at each cell level and in the erection room. Electric lighting and power systems are rated for operation in a fuel vapor environment. Emergency eyewash and showers are provided at all fuel service levels. Personnel and equipment access to the service cells is provided via "clean" and "dirty" elevators.

The administrative section of the building has 7074 sq m on two floors, consisting of offices, equipment storage, cleanroom garment changing and storage areas, and controlled personnel access. Both electronic and physical security capabilities exist for personnel access.

The basement of the facility provides high-pressure air, HVAC chillers and blowers, vacuum systems, and cable distribution to other areas of SLC-6.

A transfer tower in this facility contains a 75 ton hoist.

Additional space is available for installation of support equipment, including equipment for emergency power. CCSI has added a launch control room within this facility.

Acronyms

30 SW	30th Space Wing
CCSI	California Commercial Spaceport Incorporated
DMSP	Defense Meteorological Satellite Program
HH	hook height
HVAC	heating, ventilation, and air-conditioning
ICBM	intercontinental ballistic missile
IOP	Integrate On Pad
IRBM	intermediate range ballistic missile
LMLV	Lockheed Martin Launch Vehicle
MST	mobile service tower
NRL	Naval Research Laboratory
PPF	payload processing facility
PCM	pulse code modulation
PCR	payload changeout room
RLCC	Remote Launch Control Center
TCLC	Titan Centaur Launch Complex
VAFB	Vandenberg Air Force Base
VHF	very high frequency
WCOOA	West Coast Offshore Operating Area
WR	Western Range
WRCC	Western Range Control Center

Points of Contact

The 30th Space Wing
30 SW/XPR
806 13th Street
Suite 3B
Vandenberg AFB, California 93437-5244
U.S.A.
E-mail: FrontDoor@vandenberg.af.mil
Voice: (805) 606-7363
Fax: (805) 606-7979

Astrotech Space Operations—Vandenberg
Bldg. 1036 Tangair Rd @ Red Rd
Vandenberg AFB, California 93437
U.S.A.
Voice: (805) 875-6400
Fax: (805) 734-2551

Spaceport Systems International, L.P.
3769-C Constellation Road
Lompoc, California 93436
U.S.A.
Voice: (805) 733-7370
Fax: (805) 733-7372

References

J. Tomei and J. Gin, "Launch Vehicle Overview" class, Sections 12 and 13, The Aerospace Institute of The Aerospace Corporation, El Segundo, CA.

U.S. Centennial of Flight Commission, "Vandenberg Air Force Base," <http://www.centennialofflight.gov/essay/SPACEFLIGHT/VAFB/SP47.htm> (16 August 2005).

Vandenberg Air Force Base (home page), <http://www.vandenberg.af.mil/> (16 August 2005).

M. Wade, ed., "Vandenberg," *Encyclopedia Astronautica*, last modification 30 March 2005, <http://www.astronautix.com/sites/vannberg.htm> (16 August 2005).

J. F. Wambolt and J. F. Kephart, "A Complete Range of Launch Activities," *Crosslink* **4** (1) (Winter 2003).

Wallops Flight Facility
• suborbital • 37.83 deg north • 75.49 deg west

Overview
Wallops Flight Facility (WFF), one of the oldest launch sites in the world, serves as NASA's principal facility for launching suborbital rockets. In the past, it has also been the site for orbital launches of the Scout vehicle and the Pegasus XL with L-1011 aircraft using WFF range support. Located on Wallops Island, Virginia, the 2506 ha (6188 acre) facility is spread out over three different parcels of land and has 84 major structures, including aircraft hangars. A research airport and aircraft fleet are located on the main base. Since its founding in 1945, the launch site has grown to include six launchpads, assembly facilities, and state-of-the-art instrumentation. WFF supports the missions of sounding rockets, aircraft, and balloons to conduct research on Earth's atmosphere and the near-Earth and space environments. To date more than 15,000 launches have been conducted at WFF.

Congress authorized funding on 1 April 1945 for the National Advisory Committee for Aeronautics (NACA) to construct a rocket launch facility at Wallops

Island, Virginia, to conduct aeronautical research using rocket-propelled vehicles. The facility, part of the Langley Aeronautical Laboratory, was initially known as the Pilotless Aircraft Research Station and was supervised by Robert R. Gilruth, who later became director of NASA's Manned Spacecraft Center. By August 1945, testing had already begun with the launch of a research rocket, the Tiamat missile. During its first few years of operations, Wallops was an important center for aerodynamic and heat-transfer research.

When NASA was founded in 1958, the facility was renamed Wallops Station, and it became an independent NASA field center. At this time, Wallops was expanded to include the Chincoteague Naval Air Station, about 11 km northwest of Wallops Island. Wallops personnel were given the task of developing components for the new U.S. civilian space program, and Wallops Station was the site of the first successful test of the Mercury space capsule. In 1975, the facility was designated as the Wallops Flight Center, and in 1982 the Wallops launch site was consolidated with the Goddard Space Flight Center and renamed the Wallops Flight Facility. In addition to serving as NASA leader for suborbital programs, WFF has expanded its user base to include commercial customers.

Besides managing and implementing suborbital flight projects, WFF responsibilities include operating and managing satellite tracking stations, researching global climate change, developing new technologies that lower the cost of space access, providing support for low-cost orbital missions, and sponsoring educational outreach programs. The NASA Sounding Rocket Program, begun in 1959, also offers support to non-NASA customers, including DOD and other government agencies, on a cost-reimbursable basis.

Launch Site Description

The complexes and facilities at WFF include launchpads, launchers, blockhouses, booster preparation and payload checkout buildings, dynamic balance equipment, a timing facility, wind-measuring devices, communications and control instrumentation, optical tracking stations, and surveillance and tracking radar units. Hazardous-materials storage is also available for rocket motors and chemicals.

Launch facilities are located on Wallops Island, while processing facilities and control centers are located on both the island and the mainland. Orbital launches from WFF use additional NASA instrumentation on Bermuda and DOD instrumentation at the Eastern Range. WFF also offers support for launch operations around the world via its mobile range equipment and instrumentation.

WFF provides several launch vehicle trajectory options. Wallops Island is oriented so that a launch azimuth of 135 deg is perpendicular to the shoreline. Launch azimuths between 90 and 160 deg can usually be accommodated depending on impact ranges. For most orbital vehicles, this means orbital inclinations between 38 and approximately 60 deg. Trajectory options outside this range of launch azimuths, including polar and sun-synchronous orbits, can be achieved by performing in-flight azimuth maneuvers.

Launch Complex

The Wallops launch range includes six launchpads, three blockhouses for launch control, and assembly buildings that support the preparation and launching of both suborbital and orbital launch systems. Customer-provided launch systems can also be accommodated. WFF includes the following launch systems.

Launch Area 1

This pad has a rail launcher that supports large suborbital sounding rocket missions. It is covered by a movable environmental shelter and is currently inactive.

Launch Area Number 2 and Blockhouse Number 2

Several types of launchers are located in this area to accommodate small to medium-size suborbital launch vehicles. Vehicles launched from this area include Nike-Tomahawk, Orion, Black Brant, and Nike-Orion sounding rockets, as well as the small Arcas and Super Loki meteorological rockets. The blockhouse has two mechanical equipment rooms and 12 office/equipment rooms encompassing a total of 831 sq m.

Launch Area Number 3 (Pads 3A and 3B)

Both of these pads have been used for launching Scout rockets, although Pad 3A is currently inactive. Pad 3B, a 20K launcher for suborbital launches, is covered by a movable shelter. It is used for some of the larger sounding rockets and vehicles, such as the Aries, that are used for special-purpose missions.

Launch Area Number 5

This is a U.S. Navy launch site for the Vandal missile, a two-stage supersonic missile used as a target missile for offshore Navy surface warship defense system tests. The dual launcher on this site allows two missiles to be configured, mounted, and launched in a near-salvo mode.

Launch Areas 0-A and 0-B

Launch Areas 0-A and 0-B are commercial sites operated by the Virginia Commercial Space Flight Authority (VCSFA) as the Virginia Space Flight Center. In December 2002, the VCSFA's launch site operator's license for the two launchpads was renewed for another five years to continue work toward the goal of developing a commercial launch capability for Virginia.

Facilities

Wallops Integrated Control Center

Wallops Integrated Control Center (WICC) centralizes control of launch range and research airport operations. It provides users with range data utilizing telemetry, radar, and weather services. The data is primarily transmitted and displayed by video. Expansion of WICC facilities was completed in 1992.

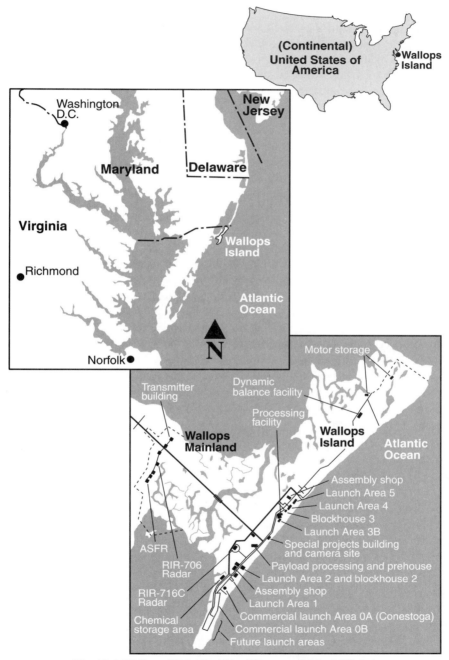

Fig. 13.6. Wallops Flight Facility. (Site map © Mark Wade.)

Storage Facilities

Liquid storage facilities are available for the temporary storage of hazardous liquids, such as propellants and purging gasses. WFF also has facilities for storing hazardous material and for the nondestructive testing of ordnance and rocket motors. Wallops has two rocket-motor storage facilities: one for Class 1.1 rocket-motor storage, and another, an aboveground facility on the north end, for storage of all classes of rocket motors.

Dynamic Balance Facility

The dynamic balance facility supports launches of sounding rockets, probes, reentries, and orbital missions. The three-building complex, located north of the launch area, includes two aerospace balancing machines. Three vertical Gisholt balancers are available for small or medium test setups with restricted diameters. Both the aerospace balancing machines and the Gisholt balancers are remotely operated and monitored from the control center.

Telemetry Systems

WFF has fixed and transportable telemetry systems/facilities to be used in support of rocket launches and low Earth orbit (LEO) spacecraft. The fixed-receiver system offers the range user a high degree of flexibility and redundancy. Each of two identical systems contains six receivers with plug-in RF heads to cover the appropriate frequency band. All systems can receive multiple links over a broad frequency spectrum. The information provided in the following sections is representative of the support capability provided by WFF for commercial launch operations. Telemetry system resources are subject to changes caused by mission requirements, revisions and modifications, and new technology.

Fixed Telemetry Systems

Multiple independently controlled telemetry antenna systems are located on the WFF main base near the approach to runway 4. These systems are controlled from the fixed-range telemetry facilities and Wallops Orbital Tracking Station (WOTS), which are collocated in building N-162. WOTS primarily supports LEO spacecraft, but its facilities are flexible enough to support range telemetry, and it shares resources with the range telemetry systems. WOTS capabilities include both metric tracking and command uplink.

Radar Systems

WFF radar systems track launch vehicles, sounding rockets, balloons, space vehicles, satellites, and aircraft to provide accurate velocity and position data. The support provided by these radar systems ranges from tracking local aircraft in the vicinity of Wallops airport to tracking distant objects in space. Radar capabilities can be enhanced by laser tracking systems and sophisticated data-processing systems to improve precision and to record, analyze, and process radar data. Radar system resources are subject to changes that result from mission requirements, revisions and modifications, and new technology.

Fixed Radar

Wallops Island radars are located just south of the Wallops Island causeway. A C-band RIR-716 radar is located at building Y-55, while the X-band Mariner's Path-finder radar is at building X-5. Mainland radars are located on the mainland just south of the causeway. These radars consist of an FPQ-6, a UHF, and an ASRF S-band SPANDAR. In addition, a C-band RIR-716 radar, which can support launch operations, is located at building A-41. Additional radar systems from the Eastern Range are not normally used in support of WFF launches; however, they can be scheduled as necessary to support mission and safety requirements.

Photooptical Systems

Still, video, and motion picture photography are available to support WFF activities and projects. Remote-controlled television cameras monitor range operations and provide safety-related information. A processing/printing laboratory and limited video editing and reproduction facilities are also available.

Tracking and Fixed Camera Stations

Tracking and fixed cameras, with media that include both film and a long-range video tracking system, provide visual information from island locations primarily for support of rocket and balloon launches.

Communications Systems

WFF operates ground-to-ground, ground-to-air, air-to-ground, ship-to-shore, range intercom, and intrastation communications systems. These systems are composed of HF/VHF/UHF radios, cables, microwave links, closed-circuit television systems, command and control communications, frequency shift tone keying systems, and high-speed data circuits. The cable plant supporting these communications systems includes extensive telephone, coaxial cable, and fiber-optic cables interconnecting the WFF facilities. Fiber-optic cables connect the main base, mainland, and Wallops Island areas. Communications provide the means for managing operations at Wallops and coordinating operations with related operations in other areas.

Communications Receiver Facility

The communications receiver facility is located on the main base in the telecommunication building, which houses the receivers, recorders, patching panels, command/destruct monitors and recorders, and supporting ancillary equipment. The receiving antennas are mounted on towers and poles in the immediate area, and worldwide reception is possible. The frequency monitoring and interference control facilities are collocated with the communications receiver facility. The communications transmitter building is located just north of, and inside, the mainland entrance to the island facility. The transmitting antennas are mounted on top of the building and on towers and poles in the immediate area. An auxiliary power generator for the redundant command/destruct and communications systems is located in an adjacent building at this facility.

Command/Destruct Systems

Ground-based command/destruct systems provide ground control of certain rocket and payload functions for range safety and/or other command purposes. The range user can use these systems to command payload functions, as necessary, within range limitations.

Fixed Command/Destruct Systems

These systems are located just north of, and inside, the mainland entrance to the island facility. Each permanent system consists of two radio transmitting sets with omnidirectional and single-helix, as well as quad-helix, antennas. Two fixed command systems are located on the mainland. These systems provide command coverage until impact, orbital insertion, or the point at which a vehicle no longer endangers the public. An electronic switch between the WFF active command site and the ER command site occurs at a predesignated time or elevation angle. This handover allows the safety officer to maintain control of the vehicle. Rockets and payloads up to those in LEO can be effectively commanded if they are within line of sight of the transmitter.

Primary Command/Destruct Subsystem

The primary command/destruct subsystem consists of an ALEPH CTS-100 transmitting set, an ANTLAB quad-helix antenna, and the necessary control circuits. The transmitter modulation can be controlled locally, or by remote control from WICC. The transmitter and the antenna pedestal operate from commercial AC power.

Redundant Command/Destruct Subsystem

An onsite generator can provide the power needed to operate a redundant command/destruct subsystem to maintain control over a rocket and its payload, should a commercial power failure occur during a mission.

Meteorological Facilities

Several meteorological facilities support launch operations at WFF. Fixed, balloon-borne, and optical sensors are available for obtaining atmospheric data. Current data from weather sensors on Wallops Island are continuously displayed on the local WFF closed-circuit TV system, and the data can be made available remotely by modem interfaces. An ionosphere sounding station provides detailed data on the ionosphere characteristics. A Dobson ozone spectrophotometer located on the mainland provides ozone measurements. Lightning-detection systems display lightning conditions both locally and nationwide, and an electric-field measurement system aids in determining the probability of lightning strikes and detecting local lightning activity.

Acronyms

LEO	low Earth orbit
VCSFA	Virginia Commercial Space Flight Authority
WFF	Wallops Flight Facility
WICC	Wallops Integrated Control Center
WOTS	Wallops Orbital Tracking Station

Point of Contact

Wallops Flight Facility
Policy and Business Relations Office
Suborbital Projects and Operations Directorate
Wallops Island, Virginia 23337
U.S.A.
Voice: (757) 824-1000

References

NASA GSFC, *Wallops Flight Facility Range User's Handbook*, Version G, 1 December 2003, <http://www.wff.nasa.gov/multimedia/docs/wffruh.pdf> (16 August 2005).

M. Wade, ed., "Wallops Island," *Encyclopedia Astronautica*, last modified 30 March 2005, <http://www.astronautix.com/sites/walsland.htm> (15 June 2005).

"Wallops Flight Facility," National Aeronautics and Space Administration Web site, last modified 3 February 2005, <http://www.wff.nasa.gov/> (15 June 2005).

14 A Multinational Site

Sea Launch
• orbital • 0 deg north • 154 deg west

Overview

Sea Launch is an international venture by American, Russian, Norwegian, and Ukrainian business partners. It is unique, not only because it is a collaborative enterprise, but also because it includes both land-based and oceangoing facilities. Sea Launch Company, LLC, was formed in 1995 to provide launch services to the commercial satellite market. Its launches are staged from a launch platform known as the *Odyssey* and controlled from Sea Launch's assembly and command ship (ACS). The *Odyssey* is a self-propelled, semisubmersible floating launch system that is based on an offshore oil-drilling platform. It rides catamaran-style on a pair of pontoons. Once it arrives at the launch location, the pontoons are submerged in order to achieve a stable launch position.

The ACS and the *Odyssey* travel from their home port of Long Beach, California, to a location some 800 km from the island of Kiritimati, along the equator. Kiritimati is one of the islands that make up the small island republic of Kiribati in the middle of the Pacific Ocean. (The English name for Kiritimati, given by Captain James Cook after he landed on Christmas Eve, 1777, is Christmas Island.) The equatorial location provides the most direct access available to geostationary orbit.

The weather in this region of the South Pacific is tropical and generally hot and humid, although moderated by trade winds. Water temperatures are uniform and major storms rarely strike Kiritimati. The region's weather is so consistent that launches are appropriate any time of year.

Sea Launch uses the heavy-lift capabilities of the *Zenit-3SL/Block DM-SL* third-stage launch vehicle to boost 4000–6000 kg payloads to a geosynchronous transfer orbit. The *Zenit-3SL* can also deliver commercial spacecraft to additional orbits, including medium Earth orbits and highly elliptical orbits. The availability of orbital options results from the absence of downrange land masses and major shipping lanes in the region.

The Boeing Company of the United States and RSC Energia, a Russian firm, are the principal corporate investors of the multinational Sea Launch Company. Boeing manufactures the payload fairing for the vehicle and provides spacecraft integration and overall mission management, while RSC Energia supplies the Block DM-SL upper stage as well as launch vehicle integration and mission operations. SDO Yuzhnoye/PO Yuzhmash (Ukraine) provides the *Zenit-3SL* first and second stages and vehicle integration support, while the Aker Kvaerner Group (Norway) is responsible for design and construction of the launch platform and command ship.

The unusual international effort that led to the creation of Sea Launch resulted from new conditions in the commercial space launch industry after the 1991 collapse of the Soviet Union. The end of Cold War rivalries allowed former adversaries to pool resources in hope of acquiring a share of what was emerging in the early 1990s to be a lucrative market for commercial launch ventures. It was logical that companies from the two space superpowers, the United States and

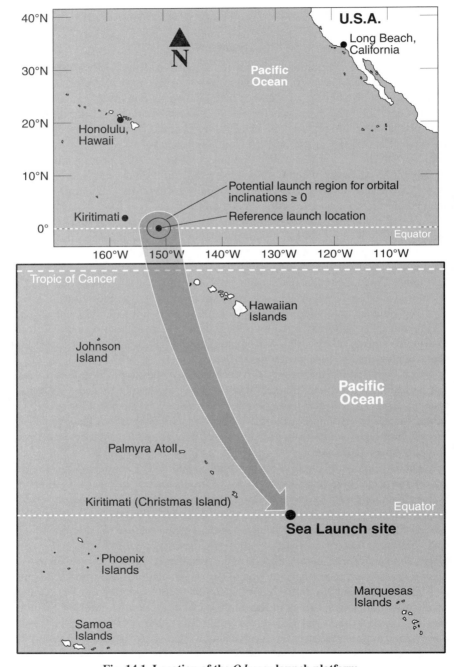

Fig. 14.1. Location of the *Odyssey* launch platform.

Russia, would participate in an attempt to exploit the commercial possibilities of space. These companies were joined by a Norwegian company that offered vital shipbuilding and offshore construction expertise, and also by a firm from the newly independent Ukraine.

During the Soviet era, Ukraine was one of the most important participants in the Soviet space program, contributing to the design and production of several missiles and spacecraft. Following the new nation's declaration of independence in 1991, a national space program was recognized as an integral part of its priorities. The National Space Agency was created to direct Ukrainian space activities, and increased commercial ventures and international cooperation were stressed as two of the agency's primary goals, which fit well with the objectives of the Sea Launch project.

Exploratory studies for the formation of Sea Launch began in summer 1993, with formal incorporation of the company following in April 1995. Construction of the command ship commenced in December 1995; construction of the Long Beach facilities began in August 1996 and was completed in 1998. During summer and fall of 1998 the company's two seagoing vessels arrived in Long Beach. The initial launch was a demonstration flight on 27 March 1999, and the first commercial launch, a DIRECTV 1-R payload, took place the following October.

Fig. 14.2. The transfer of the *Zenit-3SL* launch vehicle from the ship to the launch platform. (Courtesy Sea Launch)

Facilities

Home Port

Satellite payloads are received at Sea Launch's home port, Long Beach. Sea Launch provides customers with logistics assistance and transportation from local airports. The company also transports encapsulated spacecraft between home port facilities. The satellite processing, fueling, and payload fairing encapsulation operations take place in the payload processing facility (PPF). Customers are assigned to one of two high-bay processing and fueling cells, both certified clean rooms, that provide for electrical testing, fueling, and final assembly prior to the encapsulation of the spacecraft into the payload fairing. Spacecraft operations are conducted from control rooms adjacent to the processing and fueling cells. An oxidizer storage room and a fuel storage room support the operations of each processing and fueling cell, and both rooms offer facilities for containment of accidental leakage of hazardous materials.

An air lock gives access to either of the two processing and fueling cells and to the encapsulation cell, which contains all the equipment for encapsulating the payload within the payload fairing. After technicians give the payload a final inspection, it is transferred to the ACS for integration with the launch vehicle.

An ordnance storage building near the PPF contains explosives used by the launch vehicle and the payload. Warehouse and storage facilities for supplies and equipment are also located adjacent to the PPF and are available to customers.

Building 4 provides office and conference space for the customer and includes a pair of rooms to remotely control or monitor PPF operations.

Assembly and Command Ship

The ACS serves as the assembly, processing, and checkout facility for the launch vehicle while in port and also contains mission control operations used for sea-based launches. The 203 m long, specially designed vessel is about 32 m wide, and it has a cruising range of 18,000 nautical miles. The ACS was assembled by the Kvaerner Group in its Glasgow, Scotland, shipyard.

After the encapsulated payload is transferred from the PPF to the ACS, it is integrated with the launch vehicle in the rocket assembly compartment, located on the ship's main deck. The integrated launch vehicle is then transported to the launch platform at the home port site.

The onboard Launch Control Center (LCC) monitors and controls all operations at the launch site, using both English- and Russian-speaking crews. Operator consoles and seating, including six console stations for customers, are situated in the LCC. All consoles provide connections with all shipboard communications systems. The launch countdown and telemetry, tracking, and control data are monitored and directed from the LCC. The ship's mission management and display system displays on the consoles information received from onboard weather instruments, the Global Positioning System, the Energia launch and flight control system, and other sources. Communications Satellite Corporation (COMSAT)

Fig. 14.3. *SL GXIII* liftoff of *Galaxy XIII/Horizons 1*, September 2003. (Courtesy Sea Launch)

links allow customers to have telephone, fax machine, and teleconferencing connections to corporate locations.

The ACS also provides customer and contractor work areas, including offices, conference rooms, and storage rooms. Ground support equipment rooms allow space for equipment to control and monitor spacecraft during operations.

The *Commander* offers living and support accommodations, including a cafeteria and movie theater, for the crew and customers during the voyages to and from the launch site. The ship can accommodate 240 people.

Launch Platform

The *Odyssey*, which has an overall length of approximately 133 m and is some 67 m wide, was refurbished in Stavanger, Norway, by the Kvaerner Group while launch and support equipment was installed at Kvaerner's Vyborg, Russia, shipyard. It is equipped with living quarters and support services for up to 68 people. The *Odyssey* contains a launchpad and all necessary systems for launch vehicle fueling, prelaunch processing, and performing launch operations. The launch platform transports the integrated launch vehicle to its enclosed launch hangar on the launch deck. Approximately three hours prior to the launch, before the propellant is loaded, the crew members are transferred to the ACS by a link bridge between the two vessels or by helicopter.

Fig. 14.4. Launch of Loral's *Telstar 14/Estrela do Sul 1* on 10 January 2004. (Courtesy Sea Launch)

Remote-controlled fueling and automatic launch systems enable launches with no personnel aboard.

Support Services

Sea Launch has well-equipped range tracking and telemetry support capabilities for use once a launch has taken place. For reception of telemetry data during the initial flight phase, the ACS utilizes a line-of-sight telemetry system that includes the Proton antenna and the S-band system through payload fairing jettison. The Tracking and Data Relay Satellite System (TDRSS) is used for later phases of flight. Additional tracking facilities, which may be used for payload missions with insufficient TDRSS coverage, are available through Energia's Moscow control center and Russian ground tracking stations.

The Proton tracking antenna on the ACS receives data from the *Zenit-3SL/ Block DM-SL* from the period just before liftoff until some 7–8 min after launch. Data are transmitted to the Energia and Yuzhnoye telemetry reception areas onboard the ACS. Data from the payload unit are also received in the ACS Boeing telemetry room from the S-band system from shortly after the time of launch until about 3 min into the flight, when payload fairing separation occurs.

When the payload fairing separates, the payload unit delivers telemetry data by means of the TDRSS, which receives data from the Zenit stages, the Block DM upper stage, and the payload unit during different flight phases. The Block DM and payload data are initially combined, while Zenit data are sent to a separate TDRSS receiver. Zenit data are received by NASA's White Sands Ground Station and routed to a Sea Launch ground station in Brewster, Washington, where they are transmitted on to the ACS. The combined Block DM and payload unit data are received by the same stations, but the data are separated in Brewster before they are transmitted to the ACS.

A weather station onboard the ACS provides on-site weather forecasting for launches. A Sea Launch meteorologist analyzes weather data from a variety of sources, including a Doppler radar system, atmospheric balloons, wave radar, and wind instruments, to provide real-time launch forecasts.

Other Sea Launch support services include photography systems onboard both the ACS and the launch platform for launch documentation and an eight-passenger Bell 230 helicopter to carry out launch platform inspections and other assignments.

Acronyms

ACS	assembly and command ship
LCC	launch control center
PPF	payload processing facility
TDRSS	Tracking and Data Relay Satellite System

Points of Contact

Sea Launch Company, LLC
One World Trade Center, Suite 950
Long Beach, California 90831
U.S.A.
Voice: (562) 499-4729

Boeing Launch Services
Voice: (714) 896-5195
Fax: (714) 372-0886
E-mail: launchservices@boeing.com

References

A. R. Curtis, ed., "Satellites Launched From Pacific Platform," *Space Today Online*, 2000, <http://www.spacetoday.org/Rockets/BoeingSeaLaunch.html> (29 March 2005).

A. Markels, "The Next Wave," *Wired News*, April 2001, <http://www.wired.com/wired/archive/9.04/sealaunch_pr.html> (29 March 2005).

Sea Launch, <http://www.sea-launch.com/> (29 March 2005).

"Sea Launch System," *S. P. Korolev Rocket & Space Corporation Energia*, <http://www.energia.ru/english/energia/sea-launch/sea-launch.html> (29 March 2005).

"Sea Launch System—Commercial Heavy-Lift Launch Services, U.S.A.," *space-technology.com*, 2005 <http://www.space-technology.com/projects/sealaunch/> (29 March 2005).

M. Wade, ed., "Kiritimati," *Encyclopedia Astronautica*, last modification 5 September 2004, <http://www.astronautix.com/sites/kirimati.htm> (29 March 2005).

Appendix I: Proposed Commercial U.S. Spaceports

The National Coalition of Spaceport States was formed in February 2001 by 14 states, united by the common goal of making sure the United States remains a major player in the developing commercial space business. Many of these states had already formulated plans for spaceports but wanted to present a united front when discussing the possibilities offered by spaceport development.

"Spaceport" is simply another term for "launch site." A spaceport typically contains launchpads and runways, as well as the infrastructure and equipment necessary for processing launch vehicles and their payloads before launch. Until recently, all U.S. spaceports were built and operated by the federal government.

Backers of these proposed spaceports hope to expand the nation's transportation system to incorporate simplified access to space. Commercial customers will constitute the majority of spaceport customers, although several proposed spaceports also plan to offer their services for U.S. Government launches.

A renewed interest in commercial spaceports has developed in recent years as the possibilities for space tourism in the near future have grown. Such possibilities particularly increased following the two successful spaceflights by Burt Rutan's SpaceShipOne in September–October 2004. Space tourism will rely on the facilities of newer spaceports with a faster launch turnaround time—user-friendly facilities that are akin to those of airports. Familiar-looking facilities could help reduce the anxiety of potential space-traveling customers.

Only a few of the spaceports listed here are licensed or have any operating capacities or infrastructure.

Alabama Spaceport

The Alabama Spaceport is planning a full-service launch facility with departure and return capability for both commercial and government customers. Current plans concentrate on developing the infrastructure for launching reusable launch vehicles (RLV) to low Earth orbit (LEO), medium Earth orbit, and geosynchronous orbit. Suborbital expendable launch vehicles will also be able to use the spaceport's facilities, which could become operational in 10 years.

The Alabama Commission on Aerospace Science and Industry and the governor's office are working together to develop the spaceport, although work on the infrastructure has not yet begun. Baldwin County, east of the city of Mobile across Mobile Bay, is the proposed location. The proposed site currently consists of open field space with basic power, water, and utilities. Future launch infrastructure development will progress as customer needs are identified.

The state of Alabama would like to lease part of the land (for the flight safety zone) and to purchase another part of it (for the launch and landing site).

References

"Existing and Proposed U.S. Spaceports," <http://64.233.161.104/
search?q=cache:v056hd90eTIJ:www.gmupolicy.net/transport2003/
Spaceport%2520Infrastructure%2520Handbook%2520PDF%252011_2002/
Part1B11_02.pdf+alabama+spaceport&hl=en> (25 July 2005).

Federal Aviation Administration Associate Administrator for Commercial Space Transportation (AST), "Launch Site Details: Alabama Spaceport," <http://ast.faa.gov/linfo_vsite/maps/detail.cfm?Fac_ID=62> (25 July 2005).

Gulf Coast Regional Spaceport

The Gulf Coast Regional Spaceport is one of three independent Texas spaceport proposals supported by the Texas Aerospace Commission. The Gulf Coast Regional Spaceport Development Corporation has proposed a spaceport on an undeveloped 4000-acre tract in Brazoria County, 80 km south of Houston.

The spaceport, which would begin operations in 10 to 15 years, would be used primarily for launches of medium- and heavy-lift commercial RLVs. The corporation is currently working on an official development plan that will determine the spaceport's necessary infrastructure, and final costs are estimated to be approximately $500 million. Local governments have invested several hundred thousand dollars in the project during the last few years in the hope of creating an estimated 7000 spaceport-related jobs.

References

M. S. Bomba, "Commercial Spaceport Development in Texas: A Review of Recent Progress," <http://64.233.161.104/search?q=cache:5GD0EIA-fJ0J:isdc2005.xisp.net/ ~kmiller/isdc_archive/fileDownload.php/ %3Flink%3DfileSelect%26file_id%3D175+gulf+coast+regional+spaceport&hl=en> (25 July 2005).

Gulf Coast Regional Spaceport Development Corporation, last modification 22 April 2005, <http://www.gulfcoastspaceport.org/> (25 July 2005).

M. Wright, "Spaceport Can Take Next Step," *The Alliance: The Economic Development Alliance for Brazoria County,* 9 March 2005, <http://www.eda-bc.com/news/archives/ release.asp?id=268> (25 July 2005).

Idaho Spaceport

In 1998 the Idaho Department of Commerce launched a study to determine the advantages of launching satellites into LEO from Idaho. The investigation disclosed the qualifications of a 71 sq km site in eastern Idaho, near the National Engineering and Environmental Laboratory, for lifting and landing RLVs, such as the X-33 VentureStar, a program of Lockheed Martin and the National Aeronautics and Space Administration (NASA). Although the VentureStar program was canceled, research concerning the possibilities of launching other RLVs from Idaho continues with the encouragement and support of NASA and the Federal Aviation Administration's Office of Commercial Space Transportation.

Idaho expects to play a role in launching and landing RLVs and is prepared to partner and facilitate development of a spaceport on land in eastern Idaho set aside by the Department of Energy (DOE) for this purpose. The spaceport infrastructure will contain launchpads and gantries, fueling facilities, payload integration facilities, telemetry and tracking facilities, range safety systems, maintenance and vehicle integration facilities, and mission control and planning facilities.

References

"Diversifying Eastern Idaho's Economic Portfolio," <http://64.233.161.104/ search?q=cache:aDN-1C6uWzEJ:www.oversight.state.id.us/ov_library/ OverviewReport2000/Economic%2520development.pdf+idaho+spaceport&hl=en> (25 July 2005).

D. M. Gray, "High Flight From the High Country," *Frontier Status Report,* 19 April 2002, <http://www.frontierstatus.com/spacepolicyarticles/199908_High_Flight.html> (25 July 2005).

Mojave Spaceport

On 29 September 2004, Burt Rutan's SpaceShipOne made history when the craft became the first private-sector rocket plane to reach outer space and return safely to Earth. It was launched from the Mojave Spaceport in eastern Kern County, California.

The Kern County government established the Mojave Airport in 1935 in Mojave, California. A few years later, with war approaching, the airport was taken over by the federal government and converted into a Marine Corps auxiliary air station. Kern County reacquired the airport in 1961 and converted the facility into the Civilian Flight Test Center. East Kern Airport owns and operates the facility, and the local government is in the process of creating a development plan for the 13.4 sq km on which the Civilian Flight Test Center is located.

The spaceport would conduct payload processing, payload integration, testing, and launch services for horizontal launches of RLVs. The Civilian Flight Test Center's infrastructure includes four rocket test sites, three runways, an air control tower, a rotor test stand, engineering facilities, and a high bay building. In June 2004, the Mojave Spaceport received a spaceport license to operate commercial launches, thus becoming the first inland spaceport to receive certification from the Federal Aviation Administration. In addition to Rutan's Scaled Composites aerospace development company, several other rocket companies are based at or near the spaceport, including XCOR Aerospace.

References

R. Piquepaille, "Welcome to the Mojave Spaceport," *Roland Piquepaille's Technology Trends,* 28 May 2004, <http://www.primidi.com/2004/05/28.html#a857> (25 July 2005).

Mojave Airport, <http://www.mojaveairport.com/> (25 July 2005).

"Mojave Spaceport," *Wikipedia: The Free Encyclopedia,* last modification 11 July 2005, <http://en.wikipedia.org/wiki/Mojave_Spaceport> (25 July 2005).

M. Wade, ed., "Mojave," *Encyclopedia Astronautica,* last modification 30 March 2005, <http://www.astronautix.com/sites/mojave.htm> (25 July 2005).

Montana Spaceport

The state of Montana established the Montana Aerospace Development Authority (MADA) to coordinate and lead Montana's commercial space efforts. The non-profit authority is governed by a nine-member board. Montana's space strategy involves creating the organizational and educational infrastructure necessary to support aerospace research and development activities and to ultimately construct a commercial spaceport. The state has authorized $20 million in aerospace bonding available to finance activities directly related to aerospace research and development or the development of spaceport infrastructure.

The proposed spaceport would launch RLVs from two possible sites: One is at Malmstrom Air Force Base in Great Falls, and the other is the deactivated Glasgow Air Force Base near the town of Glasgow. In 2000, MADA actively worked with officials from both Lockheed Martin's X-33 VentureStar space plane project and the Rotary Rocket Company in an attempt to bring commercial space launches to the state, but following the cancellation of the VentureStar program in 2001, the licensing process was put on hold. Despite an initially aggressive effort by Montana officials to develop a spaceport, its future is uncertain and planning is on hold.

References

Federal Aviation Administration Associate Administrator for Commercial Space Transportation (AST), "Launch Site Details: Montana Spaceport," <http://ast.faa.gov/linfo_vsite/maps/detail.cfm?Fac_ID=50> (25 July 2005).

1999 Montana Legislature, House Bill No. 555, 31 March 1999, <http://ssl.csg.org/dockets/20cycle/2000c/20cbills/0420C01mt.htm> (25 July 2005).

Nevada Test Site

The Nevada Test Site, located 100 km northwest of Las Vegas, is a remote, highly secure facility that was formerly the site of most U.S. nuclear tests, both below-ground and aboveground. Kistler Aerospace Corporation selected the site as a potential spaceport for its K-1 RLV in addition to its Woomera, Australia, facility to increase scheduling flexibility and to widen the range of available launch azimuths. Lockheed Martin has also expressed interest in using the site.

Although it does not currently have any launch infrastructure, the Nevada Test Site has basic infrastructure (such as a paved runway, roads, water, and power) that can be used to support launch-related activities. The Nevada Test Site

Development Corporation (NTSDC) obtained an economic development use permit in 1997 from the U.S. DOE. The permit allowed the corporation to pursue development of the Nevada Test Site as a spaceport. The NTSDC later issued a permit allowing Kistler to operate a launch and recovery operation at the Nevada Test Site.

References

M. Wade, ed., "Nevada Test Site," *Encyclopedia Astronautica*, last modification 30 March 2005, <http://www.astronautix.com/sites/nevtsite.htm> (25 July 2005).

Kistler Aerospace Corporation (home page), 2005, <http://www.kistleraerospace.com/> (25 July 2005).

"Kistler Aerospace Corporation's K-1," *GlobalSecurity.org*, last modification 27 April 2005, <http://www.globalsecurity.org/space/systems/kistler.htm> (25 July 2005).

V. J. Brechin and G. Smith, "Space Cowboys," *Ye Olde Consciousness Shoppe*, 15 January 2001, <http://yeoldeconsciousnessshoppe.com/art35.html> (25 July 2005).

Oklahoma Spaceport

The state of Oklahoma has actively worked to develop both a broader space-related industrial base and a commercial spaceport. In May 1999, Governor Frank Keating signed a bill authorizing the creation of the Oklahoma Space Industry Development Authority (OSIDA). Its mission is to promote the development of spaceport facilities, space exploration, space education, and aerospace industries in Oklahoma. The same month, the governor also signed the Space Industry Tax Incentive Act, which offered tax credits, tax exemptions, and accelerated depreciation rates for commercial spaceport-related activities. The deactivated Clinton-Sherman Air Force Base at Burns Flat was selected as the site of the Oklahoma Spaceport, which was formally inaugurated in March 2002.

The existing infrastructure includes a 4100 m runway originally built to service B-52 aircraft, six large hangars, a rail spur, utilities, and 12.4 sq km of undeveloped land. Proposals have also been made to build a rocket engine manufacturing plant and training facilities adjacent to the spaceport. The Oklahoma Spaceport will be limited to launch and support services for RLVs, because with an inland location, spent rocket stages would fall on populated areas. A study for the now-canceled VentureStar space plane concluded that Oklahoma could be used for launching RLVs into polar orbits.

OSIDA has signed a memorandum of understanding with Rocketplane Ltd., along with several other companies, for use of the Burns Flat site. Rocketplane Ltd. qualified for tax credits from the state in 2003. Launches at the Oklahoma Spaceport may take place by 2007, while OSIDA's goal is to have the spaceport fully operational by 2025.

References

Oklahoma Space Industry Development Authority, 2005,
<http://www.okspaceport.state.ok.us/> (25 July 2005).

J. Foust, "Little Spaceport on the Prairie," *The Space Review*, 7 June 2004,
<http://www.thespacereview.com/article/157/1> (25 July 2005).

L. David, "Oklahoma Steps Up To Space," *SPACE.com*, 10 December 2001, <http://
www.space.com/missionlaunches/ok_spaceport_011210.html> (25 July 2005).

South Dakota Spaceport

The South Dakota state government has been involved in preliminary planning
and feasibility studies regarding the viability of establishing a spaceport in the
state as a result of inquiries from both the private sector and NASA. The state has
selected a possible site near Ellsworth Air Force Base, located close to Rapid City.

Although the state of South Dakota may ultimately own a portion of the proposed spaceport, the spaceport's operating entity has not yet been established.
Planning has so far involved a mixture of local and state government officials as
well as National Guard members.

Currently, no infrastructure is available, and future construction of any infrastructure and the size of the spaceport's site will depend on customer needs. However, any funding is contingent on the state's receiving a written commitment from
a private company or government organization to host an RLV program. The state
government plans to provide several other incentives for the spaceport, including
bonds, tax deductions, and investment and job training credits.

References

Federal Aviation Administration Associate Administrator for Commercial Space
Transportation (AST), "Launch Site Details: South Dakota Spaceport," <http://ast.faa.gov/
linfo_vsite/maps/detail.cfm?Fac_ID=59> (25 July 2005).

B. Harlan, "A 'Galactic' Proposal for a Virgin Mogul," *Rapid City Journal*, 2005, <http://
www.rapidcityjournal.com/articles/2005/07/11/news/columns/214harlan.prt> (25 July
2005).

South Texas Spaceport

The South Texas Spaceport is one of the three Texas spaceport proposals being
supported by the Texas Aerospace Commission. In December 2001, the Willacy
County Commissioners Court established the Willacy County Development Corporation for Spaceport Facilities. The corporation is investigating the feasibility of
developing a spaceport on a 40 sq km parcel of unused land in Willacy County on
the coast of the Gulf of Mexico, 150 km south of Corpus Christi.

The spaceport would primarily support commercial RLVs, although some expendable orbital and suborbital rockets could also use it. The Willacy County project replaces an earlier proposal for a spaceport in neighboring Kenedy County, 65 km to the north, that was met by strong opposition from county residents.

References

M. S. Bomba, "Commercial Spaceport Development in Texas: A Review of Recent Progress," <http://64.233.161.104/search?q=cache:5GD0EIA-fJ0J:isdc2005.xisp.net/ ~kmiller/isdc_archive/fileDownload.php/ %3Flink%3DfileSelect%26file_id%3D175+gulf+coast+regional+spaceport&hl=en> (25 July 2005).

M. Drosjack, "Former Astronaut Pushes Legislature for Spaceport Boost," *Houston Chronicle*, 25 February 2003, <http://www.chron.com/cs/CDA/story.hts/space/1795173> (25 July 2005).

Southwest Regional Spaceport

New Mexico plans to construct and operate the Southwest Regional Spaceport for use by both commercial and government customers conducting space activities and operations. The Southwest Regional Spaceport is supported by the state through the New Mexico Office of Space Commercialization (NMOSC), established in 1994 as part of the New Mexico Economic Development Department. The goal of NMOSC is to coordinate the marketing and promotion of the state's space-related resources.

The proposed site of the spaceport is a 43 sq km parcel of open land, approximately 72 km north of Las Cruces, near the town of Upham, in the south-central part of the state. The spaceport will support launches of all classes of RLVs and provide support for payload integration and vehicle landings. The facility will accommodate horizontal and vertical launches and horizontal landings and will include a launch complex, a 3600 m runway and aviation complex, a payload assembly complex, and a support facilities complex.

In 2001, the state legislature approved $1.5 million in funds for fiscal years 2002 through 2004 for spaceport development, including environmental studies and land acquisition. The funding is contingent on New Mexico receiving a written commitment from a commercial or government customer to utilize the Southwest Regional Spaceport for a designated RLV program. The state legislature has provided several other incentives for the spaceport, including tax deductions, bonds, and investment and job training credits.

In 2004, the Southwest Regional Spaceport was chosen by the X Prize Foundation to be the site of the X Prize Cup, which will be an annual competition for RLVs. The first events, including demonstration flights, will take place in October 2005, with the first rocket races scheduled for the fall of 2006.

References
K. Boehler, "X Marks the Spot," *enchantment*, 2005,
<http://www.enchantment.coop/features/0507space.php> (25 July 2005).

"New Mexico Office for Space Commercialization," *New Mexico Economic Development Department*, 2004, <http://testweb.edd.state.nm.us/SPACE/SPACEPORT/space.html> (25 July 2005).

L. David, "New Mexico to Host Private 'X Prize' Spaceflight Race," *SPACE.com*, 11 May 2004, <http://www.space.com/news/xprize_newmexico_040511.html> (25 July 2005).

"Governor Bill Richardson Announces X Prize Cup Events," *Space Race News*, 13 April 2005, <http://www.xprizenews.org/index.php?p=867> (25 July 2005).

Spaceport Washington

Spaceport Washington, a joint public and private partnership, has selected Grant County International Airport in central Washington, 280 km east of Seattle, as a possible spaceport site. It is located in a low-density population area, on flat, open land that has access to both highway and rail transportation. The airport, which is on the site of the former Larson Air Force Base and is now owned and operated by the Port of Moses Lake, is used largely as a testing and training facility, and it has been certified as an emergency landing site for the space shuttle.

Spaceport Washington will use the airport for horizontal and vertical takeoffs and horizontal landings of RLVs. The now-canceled Lockheed Martin VentureStar craft was once suggested as a potential customer. Available infrastructure at the airport includes a 4100 m main runway and a 3200 m crosswind runway. In addition, Grant County has an established industrial infrastructure capable of supporting future launches. Spaceport Washington has received both funding and staff support from the state of Washington.

References
ASPI Group, "Spaceport Washington," 2005, <http://www.aspigroup.com/properties_indu_induspace/induspace.htm> (25 July 2005).

Federal Aviation Administration Associate Administrator for Commercial Space Transportation (AST), "Launch Site Details: Washington Spaceport," <http://ast.faa.gov/linfo_vsite/maps/detail.cfm?Fac_ID=58> (25 July 2005).

Spaceport Washington, <http://www.spaceportwashington.com/> (25 July 2005).

Utah Spaceport

The state legislature of Utah passed the Utah Spaceport Authority Act in 2001. The act gave the Utah Spaceport Authority the power to develop and regulate spaceport facilities in the state. It also created a seven-member Spaceport Advisory Board appointed by the governor to advise the Authority on spaceport issues and establish regulations for the proposed spaceport. The Wah Wah Valley Interlocal Cooperation Entity proposed to construct and operate a commercial launch site using approximately 280 sq km of Utah state trust lands located 55 km southwest of Milford, Utah, in Beaver County.

The mission of the Utah Spaceport is to provide a cost-effective launch and recovery facility for RLVs. No infrastructure exists, but the proposed spaceport would include a new 4575 m runway and two vehicle launch facilities. Assembly, processing, and testing facilities are also planned. The state of Utah allocated $300,000 to conduct a spaceport feasibility study and asked the advisory board to research the economic development opportunities of the X-33 and other RLVs. The study was placed on hold following the cancellation of the X-33 and Venture-Star programs.

References

J. Bauman, "Utah Has Future as Spaceport," *Deseret Morning News*, 5 January 2004, <http://deseretnews.com/dn/view/0,1249,575041346,00.html> (25 July 2005).

Federal Aviation Administration Associate Administrator for Commercial Space Transportation (AST), "Launch Site Details: Utah Spaceport," <http://ast.faa.gov/linfo_vsite/maps/detail.cfm?Fac_ID=51> (25 July 2005).

"Governor Names Deputy Director of SDL to Spaceport Board," *Space Dynamics Laboratory*, 25 July 2001, <http://www.sdl.usu.edu/news/press/2001/employees/redd-spaceport-board> (25 July 2005).

West Texas Spaceport

The West Texas Spaceport is one of the three Texas spaceport proposals supported by the Texas Aerospace Commission. The Pecos County/West Texas Spaceport Development Corporation, established in 2001, has proposed the development of a spaceport 16 km south of Fort Stockton, Texas, on treeless, barren ranch land. The spaceport will serve vertical takeoff and landing of RLVs, with a particular emphasis on Kistler Aerospace Corporation's K-1 vehicle, as well as suborbital sounding rockets. When fully operational, the spaceport will include a launch site with a 4570 m safety radius, an adjacent recovery zone, and payload integration and launch control facilities. In October 2002, JP Aerospace, a California-based company, successfully launched a 4.2 m rocket to an altitude of 5181 m to mark the official opening of the West Texas Spaceport.

References

M. S. Bomba, "Commercial Spaceport Development in Texas: A Review of Recent Progress," <http://64.233.161.104/search?q=cache:5GD0EIA-fJ0J:isdc2005.xisp.net/~kmiller/isdc_archive/fileDownload.php/

%3Flink%3DfileSelect%26file_id%3D175+gulf+coast+regional+spaceport&hl=en> (25 July 2005).

J. Tedesco, "Successful Rocket Launch Is Aimed at Small Satellite Niche," *GlobalSecurity.org*, 6 October 2002, <http://www.globalsecurity.org/org/news/2002/021006-space1.htm> (25 July 2005).

E. Todd, "Fort Stockton's 'Spaceport': West Texas Pioneer in Aerospace," *MyWestTexas.com*, 26 August 2002, <http://www.mywesttexas.com/site/news.cfm?newsid=5157657&BRD=2288&PAG=461&dept_id=475626&rfi=6> (25 July 2005).

Wisconsin Spaceport

The Wisconsin Department of Transportation officially approved the creation of the proposed Wisconsin Spaceport in August 2000 for launchings of commercial RLVs. A site on Lake Michigan, owned by the city of Sheboygan, was selected. A spaceport for launching sounding rockets had been operational there since 1995, when the first launch occurred.

The goal of the existing spaceport, Spaceport Sheboygan, is to support space research and education through suborbital launches of sounding rockets for student projects. Rockets for Schools, a program run by Space Explorers, Inc., operates the student program, which allows students to design their own payloads and participate in launches.

While suborbital sounding rocket launches to altitudes of up to 55 km have been conducted to date, future plans for the proposed Wisconsin Spaceport include the capability for conducting orbital launches of RLVs from the Sheboygan site. The existing infrastructure includes a vertical pad for suborbital launches in addition to portable launch facilities, such as mission control, which are put up and taken down as needed. The development of RLV launch infrastructure for the Wisconsin Spaceport is being planned.

References

"Rockets for Schools: Spaceport Sheboygan," 2004, <http://www.rockets4schools.org/> (25 July 2005).

W. G. Foerster, "What's Up: A List of Projects We Are Supporting," *Sheboygan Space Society*, last modification 12 June 2004, <http://www.tcei.com/sss/SSSproj.html> (25 July 2005).

Appendix II: Points of Contact Quick Reference

Entity, Function, or Contact	Mailing Address	Phone (Voice)	Phone (Fax)	Email Address
Australia				
Area Administrator Woomera	Defence Support Centre Woomera Dewrang Avenue Woomera, South Australia 5720 Australia	+61 8 8674 3201	+61 8 8674 3308	aaw@dfence.gov.au
Commander ARDU	Aircraft Research and Development Unit RAAF Base Edinburgh Edinburgh, South Australia 5111 Australia	+61 8 8393 2111	+61 8 8393 2498	
Space Policy Unit	Department of Industry, Science and Resources P.O. Box 9839 Canberra, Australian Capital Territory 260 Australia	+61 6 213 7246	+61 6 213 7249 Telex: AA62654	egrohovaz@isr.gov.au
Brazil				
General coordination	Departamento de Pesquisas y Desenvolvimento Esplanada dos Ministérios– Bloco M Edifício Anexo do MAer, 3° Andar 70.045-900–Brasília–DF Brasil		+55 61 224 6123	

Entity, Function, or Contact	Mailing Address	Phone (Voice)	Phone (Fax)	Email Address
Technical and operational information	Centro de Lançamento de Alcântara CEP: 65.250-000–Alcântara–MA Brasil		+55 98 211 1069	
Centro de Lançamento da Barreira do Inferno		+55 84 211 0945	+55 84 211 4226	
Canada				
Churchill Rocket & Research Centre Committee	Room 648-155 Carlton Street Winnipeg, Manitoba Canada R3C 3H8	(204) 945-8193	(204) 945-8229	
China				
China Great Wall Industry Corporation	30 Haidian Nanlu, Beijing 100080, P.R.China	86-10-68748888/ 68748810	86-10-68748876/ 68748865	cgwic@cgwic.com Web: www.cgwic.com
G W Aerospace Inc.	21515 Hawthorne Blvd Suite 1065 Torrance, CA 90503-6518 U.S.A.	(310) 540-7706		
France and French Guiana				
Arianespace Inc.	601 13th St. NW Washington, D.C. 20005 U.S.A.	(202) 628-3936	(202) 628-3949	
Arianespace	Boulevard de l'Europe 91006 Evry Courcouronnes Cedex, France	+33-1-60-87-60-00	+33-1-60-87-63-04	

Entity, Function, or Contact	Mailing Address	Phone (Voice)	Phone (Fax)	Email Address
India				
Antrix Complex	New BEL Road, Bangalore 560 094 India	+91-80-3415474	+91-80-3418981	
Indian Space Research Organization	New BEL Road, Bangalore 560 094 India	+91-80-3415275 & 3415474	+91-80-3412253	
Vikram Sarabhai Space Center	Trivandrum, 695022 Kerala, India	+91-471 562-444/562-555	+91-471-7979	
Sriharikota Launch Range	Andhra Pradesh 524124 India	+91-2001-041-394	+91-2001-041-568-594	
Israel				
Israel Space Agency	P.O. Box 17185 26a Chaim Levanon Street Ramat-Aviv, 61171 Tel Aviv, Israel	(+972) 3 216 852	(+972) 3 642 2298	
Japan				
Uchinoura Space Center	1791-13 Minamikata Uchinoura-cho Kimotsuki-gun Kagoshima 893-1402 Japan	81-994-31-6978	81-994-67-3811	
Tanegashima Space Center	Mazu, Kukinaga Minamitane-machi, Kumage-gun Kagoshima 891-3793 Japan	81-9972-6-2111	81-9972-4-4004	

(Kazakhstan follows Russia)

Entity, Function, or Contact	Mailing Address	Phone (Voice)	Phone (Fax)	Email Address
Norway				
Norsk Romsenter (Norwegian Space Centre)	Hoffsveien 65A P.O. Box 85 Smestad N-0309 Oslo 9 Norway	+47 2 52 38 00	+47 2 23 97	
Andøya Rocket Range	P.O. Box 54 N-8483 Andenes Norway	+47 76 14 44 00	+47 76 14 44 01	
Hallstein Thommassen Andøya Rocket Range	P.O. Box 54 N-8480 Andenes Norway	+47 76 14 16 44		
Pakistan				
Pakistan Space and Upper Atmosphere Research Commission (SUPARCO)	Sector 28, Gulzar-e-Hijri P.O. Box 8402 Karachi 75270 Pakistan	(92-21) 8144667-74, 8144923-927	(92-21) 8144928, 8144941	suparco@ digicom.net.pk
Russia				
Plesetsk Space Center	Mirniy-12, Arkhangelskaya Oblast, 164170 Russia	+7 (095) 330-9190	+7 (095) 330-9190	
Design Bureau of Transport Machinery (KBTM)	101 Vernadsky Prospect, Bldg. 2 Moscow 117415 Russia	+7 (095) 433-3239	+7 (095) 433-1548	kbtmto@dol.ru
Cosmos International Satellitenstart GmbH	Universitätsallee 29, 28359 Bremen Germany	+49 (0) 421 2020-8	+49 (0) 421 2020-700	cosmos@ fuchs-gruppe.com

Entity, Function, or Contact	Mailing Address	Phone (Voice)	Phone (Fax)	Email Address
Eurockot Launch Services GmbH	P.O. Box 28 61 46 D-28361 Bremen Germany	+49 421 539-65 01	+49 421 539-65 00	eurockot@astrium-space.com
Starsem	2, rue François Truffaut 91042 Evry Cedex France	+33 1-69-87-01-10	+33 1-60-78-31-99	
United Start Corporation	2995 Airway Avenue Costa Mesa, CA 92626 U.S.A.	714-755-7427	714-545-7676	info@unitedstart.com
(Kazakhstan)				
Proton International Launch Services	1660 International Drive Suite 800 McLean, Virginia 22102 U.S.A.	571-633-7400	571-633-7500	
Soyuz Starsem	5-7, rue François Truffaut 91042 Evry Cedex France	+33 1- 698 70110	+33 1-607- 83199	
Rockot Eurorockot Launch Services GmbH	P.O. Box 28 61 46 Airport Center Flughafenallee 26 D-28199 Bremen Germany	+49 421 539-65 01	+49 421 539-65 00	eurockot@astrium-space.com
Zenit Yangel Yuzhnoye State Design Office	49008 Dnipropetrovsk Krivorizka str., 3 Ukraine	38 (0562) 42 00 22	38 (056) 770 01 25; 38 (0562) 92 50 41; Teletype: 143294 deviz	info@yuzhnoye.com

Entity, Function, or Contact	Mailing Address	Phone (Voice)	Phone (Fax)	Email Address
Tsyklon-2 United Start Corporation	2995 Airway Avenue Costa Mesa, California 92626 U.S.A.	(714) 755-7427	714-545-7676	info@unitedstart.com
Sweden				
Swedish National Space Board	Solna strandväg 86 Box 4006, 17104 Solna Sweden	+46 8 6276480	+46 8 6275014	spaceboard@snsb.se
Swedish Space Corporation, Esrange	P.O. Box 802 SE-981 28 Kiruna Sweden	+46 980 720 00		
United States				
45th Space Wing/XP	Patrick Air Force Base, Florida 32925 U.S.A.	(321) 494-8869		spacebiz@ patrick.af.mil
The 30th Space Wing	30 SW/XPR 806 13th Street Suite 3B Vandenberg AFB, California 93437-5244 U.S.A.	(805) 606-7363	(805) 606-7979	FrontDoor@ vandenberg.af.mil
Astrotech Space Operations—Vandenberg	Bldg. 1036 Tangair Rd @ Red Rd Vandenberg AFB, California 93437 U.S.A.	(805) 875-6400	(805) 734-2551	

Entity, Function, or Contact	Mailing Address	Phone (Voice)	Phone (Fax)	Email Address
Spaceport Systems International, L.P.	3769-C Constellation Road Lompoc, California 93436 U.S.A.	(805) 733-7370	(805) 733-7372	
Wallops Flight Facility	Policy and Business Relations Office Suborbital Projects and Operations Directorate Wallops Island, Virginia 23337 U.S.A.	(757) 824-1000		
(Sea Launch)				
Sea Launch Company, LLC	One World Trade Center, Suite 950 Long Beach, California 90831 U.S.A.	(562) 499-4729		
Boeing Launch Services		(714) 896-5195	(714) 372-0886	launchservices@ boeing.com

Index

A

Adelaide, Australia, 1
  connection to Woomera, 4
Advanced Ballistic Reentry Systems program, 118
Aerobee (sounding rocket), 21, 23, 24
AeroLab Development Co., 23
Aerospace Corporation. *See* The Aerospace Corporation
Africa, 108, 110
Air Research and Development Command, 106
Aker Kvaerner Group, 139–141, 142, 144
Akita (Japan), 57
Alabama Spaceport, 147
Alaska, 104
Alaska Aerospace Development Corporation, 104
Alcântara Launch Center, 9, 11–14
  CLBI services for, 16
  technical facilities of, 13–14
Algeria, 39
Allied Signal Corporation, 126
ALOMAR. *See* Arctic Lidar Observatory for Middle Atmosphere Research
Alouette I (satellite), 21
Anchorage, Alaska, 104, 120
Andenes, Norway, 69
Anderson Peak (California), 124, 125
Andøya Air Force Base (Norway), 69
Andøya Island (Norway), 69
Andøya Rocket Range (ARR), Norway, 69–71
Angara (launch vehicle), 79, 81
Antigua, 108, 109, 110, 115
Antigua Air Station, 108, 110, 115
Apache (launch vehicle). *See under* Nike
Apollo program, 77–78, 103, 106
Apollo-Soyuz Test Project, 103
Arcas (meteorological rocket), 132
Archangel, Russia, 79, 81
Arctic Circle, 95
Arctic Lidar Observatory for Middle Atmosphere Research (ALOMAR) Research Station, 71
Area Administrator Woomera, 3
Arena Arctica, 98
Argentia, Newfoundland, 108, 114–115
Ariane (launch vehicle), 16, 41, 42, 44
  Ariane 1, 41, 42
  Ariane 2, 42
  Ariane 3, 42
  Ariane 4, 41, 42
  Ariane 5, 41, 42, 44, 45
Arianespace Company, 41
Aries (launch vehicle), 132
ARR. *See* Andøya Rocket Range
Aryabhata (satellite), 47
Ascension, 108, 109, 110, 115
Ascension Auxiliary Air Field, 108, 115–116
Asiasat 1 (satellite), 31
Asterix 1 (satellite), 39
Astrotech Space Operations, payload processing
  facility of, 127
Athena (launch vehicle), 114, 120
Atlantic Ocean, 39, 41, 108, 110, 115
Atlas (launch vehicle), 112, 118, 120
  Atlas D/Agena, 122
  Atlas E, 122
  Atlas E/F/H, 122
  Atlas F, 118
  Atlas I, 108, 122
  Atlas II, 108, 113
  Atlas III, 108, 113
  Atlas V, 108, 109, 114, 121, 122, 124
  Atlas/Agena, 122
  Atlas/Centaur, 113, 122
Australia, 1–7, 150
Australia Defence Forces, 1
Australian Department of Defence, 3
Australian Space Council, 3

B

B-52 aircraft, 151
Badar (satellite), 73
Baikonur Cosmodrome, 73, 77–79, 81, 85–92
Balasore, India, 49
Banana River, 108, 110, 112
Banana River Naval Air Station, 106
Barreira do Inferno Sounding Rocket Range, 9, 14–18
  See also Centro de Lançamento da Barreira do Inferno
Beijing, China, 31
Bermuda, 108, 131
Black Brant (sounding rocket), 3, 15, 23–24, 69, 95–97, 132
Black Knight (rocket), launch of, from Woomera, 1
Blue Streak missiles, tests of, 4
Boeing 747 (aircraft), 4, 98
  at Xichang airport, 31
Boeing. *See* The Boeing Company
Boso peninsula, 65
Brazil, 9–19
Brazilian Space Agency, 9
Brazilian Telecommunications Company, 17
Brewster, Washington, 145
British West Indies, 110, 115
Bush, President George W., 104

C

C-130 (aircraft), at Xichang airport, 31
California, 104, 128, 149, 155
  coast of, 120–121
  home port of Sea Launch in, 139
  location of VAFB in, 117
  See also Kern County; Lompoc; Long Beach; Los Angeles; San Diego; San Francisco; Santa Barbara County; Santa Maria

California Commercial Spaceport Incorporated (CCSI), 127, 128
Camp Cooke, 117
Canada, 21–25
  army of, 21
  Eastern Range support in, 108, 114
Canadian Launch Safety Office, 22
Canadian Space Agency, 21
Cape Canaveral Air Station (CCAS), 82, 105–116, 120
Caribbean Sea, 108, 110
Castor IV A, 122
Castor Lance (sounding rocket), 15
Cayenne, French Guiana, 39
CCAFS. See Cape Canaveral Air Force Station
CCSI. See California Commercial Spaceport Incorporated
Centaur (launch vehicle)
  in program with Atlas
    at CCAS, 113
    at VAFB, 122
  in program with Titan IV
    at CCAS, 114
    at VAFB, 123
Centaure (rocket), 69, 73, 95
Centre National d'Études Spatiales (CNES), 39–45, 69
Centre Spatial Guyanais (CSG), 39–45
Centre Spatiale de Toulouse, 16
Centro de Lançamento da Barreira do Inferno (CLBI), 15
  technical facilities of, 16–18
  user support services offered by, 18
  See also Barreira do Inferno Sounding Rocket Range
Challenger space shuttle, 57, 103, 108, 118, 123
Changzheng. See Long March
China, 27–36
  successful astronaut launch by, 47, 59
  use of Alcântara (Brazil) launch site by, 11
China Great Wall Industry Corporation, 27
China Satellite Launch and Tracking Control General (CLTC), 31
Chincoteague Naval Air Station, 131
Christmas Island. See Kiritimati
Churchill Launch Range. See Churchill Rocket & Research Range
Churchill Northern Studies Centre, 21
Churchill Rocket & Research Centre Committee, 21
Churchill Rocket & Research Centre. See Churchill Rocket & Research Range
Churchill Rocket & Research Range, 21–25
Churchill, Manitoba, 21
Clark, Massachusetts, 101
Clementine Project, 104
CLTC. See China Satellite Launch and Tracking Control General

CNES. See Centre National d'Études Spatiales
Cold War, 1, 77, 101–102, 139
Columbia space shuttle, 53, 79, 103–104, 106
commercial space activities
  Australian, 3
  Brazilian, 9–10, 11
  Canadian, 21
  Chinese, 27, 28, 31, 33, 34, 36
  French, 39, 41, 42
  Indian, 47
  Israeli, 55
  Kazakh, 87, 90
  provided by Sea Launch, 139–141
  Russian, 82, 87
  U.S., 105, 113, 114, 121, 122, 124, 125, 127–128, 131, 134, 147–156
Communications Satellite Corporation (COMSAT), 142
Complete Brazilian Space Mission, 9
COMSAT. See Communications Satellite Corporation
Cook, Capt. James, 139
Cooke, Maj. Gen. Philip St. George, 117
Corona (satellites), 103
Cosmos 112 (satellite), 81
Crétien, Jean-Loup, 39
crewed spaceflight, 1, 27, 28, 30, 47, 59, 78, 85, 106, 118
CSG. See Centre Spatial Guyanais
Cyclone-4 (rocket), 11

D

D-8 (launcher), 23
Dart (sounding rocket), 69
de Gaulle, President Charles, of France, 39
Defence Support Centre Woomera (DSCW), 3
Defense Meteorological Satellite Program (DMSP), 103, 123
  payload test facility of, 127
Delta (launch vehicle), 57, 112, 113, 118, 121–122
  booster processing of at Western Range, 124
  Delta II, 109, 113, 121, 124, 126
  Delta III, 108, 113
  Delta IV, 108, 109, 114, 120, 121, 123, 124
  Delta IV Evolved Expendable Launch Vehicle program, 120
DF series missiles, 27, 34
DFH-1 (satellite), 27
Dhawan, Satish, 49
  See also Satish Dhawan Space Center
Diamant (launch vehicle), 39, 41, 42
DIRECTV, 1-R payload of, on Sea Launch, 141
Discoverer I (satellite), 118
Discovery space shuttle, 103–104
DMSP. See Defense Meteorological Satellite Program

Dnepr (rocket), 85, 87–88
docking techniques, 77
DOD. *See* U.S. Department of Defense
DOE. *See* U.S. Department of Energy, 151
Dong Feng base area within JSLC, 30
Dragon (rocket), 69, 73
DSCW. *See* Defence Support Centre Woomera
Dyna-Soar program, 101, 103, 106

E

EADS Space Transportation, 41
Earth Observing System, Terra satellite of, 122
EASP. *See* Esrange Andøya Special Project
Eastern Range (United States), 104, 105–116,
    120, 131, 135
EELV. *See* evolved expendable launch vehicle
EGIS. *See* Esrange Geophysical Information
    Services
Eisenhower, President Dwight D., 101
ELA launchpad. *See* Ensemble de Lancement
    Ariane launchpad
ELDO. *See* European Launcher Development
    Organization
ELV. *See* expendable launch vehicle
Energia (Russian company). *See* RSC Energia
Ensemble de Lancement Ariane (ELA) launch-
    pad, 41–44
Enterprise (shuttle flight verification vehicle),
    118
equatorial orbit, 11
ESA. *See* European Space Agency
ESP. *See* Esrange Special Project
Esrange Andøya Special Project (EASP), 95
Esrange Division (ED), 97
Esrange Geophysical Information Services
    (EGIS) system, 97
Esrange launch site, 69, 95–98
    rocket impact area of, 97
    Satellite Operations Division of, 97
Esrange Special Project (ESP), 95
ESRO. *See* European Space Research Organiza-
    tion
Etsin River (China), 30
Europa (rocket)
    Europa II, 41–42
    launch of, from Woomera, 1
European Launcher Development Organization
    (ELDO), 1
European Space Agency (ESA), 15, 39–41, 42,
    53, 69, 95
European Space Conference, 41
European Space Research Organization (ESRO),
    69, 95
Eutelsat, 87
evolved expendable launch vehicle (EELV), 108
    Atlas V, 114
    Delta IV, 114

expendable launch vehicle (ELV), 106–109, 147
Explorer (satellite), 77, 102
    Explorer 1, 101, 106
export controls, effect of, on U.S. use of Chinese
    facilities, 27

F

Fairbanks, Alaska, 104
Falcon I (launch vehicle), 121, 122
Federal Aviation Administration, 120
    Office of Commercial Space Transporta-
        tion, 148
Federal Communications Commission, 124
Fenhe River (China), 34
Finland, coverage of, by impact area of Esrange,
    97
Flight Test Range (FTR) of Pakistan, 73–75
Florida, 104
    coast of, instrumentation sites along, 108
    consolidation of shuttle operations in, 120
    in ER oversight area, 105
    scheduled landing of Columbia in, 53
    See also Banana River; Florida Space Au-
        thority; Jacksonville; Jonathan
        Dickinson Park; Merritt Island;
        Miami; Palm Bay; Tequesta
Florida Space Authority, 113, 114
Fort Churchill, 21
France, 39–45
    Centre Spatiale de Toulouse of, support by,
        for CLBI, 16
    and launch of commercial Israeli satellites,
        55
    launch of satellites into orbit by, 27
    membership of, in EASP/ESA, 95
Freedom 7, 106
French Guiana, 11, 16, 39–45
FTR. *See* Flight Test Range

G

Gagarin, Yuri, 77–78, 85, 102
Gansu Province (China), 30
Gemini program, 77, 103, 106
Geosynchronous Launch Vehicle (GSLV), 47–
    49
Germany
    Institute for Atmospheric Physics of, 71
    membership of, in EASP/ESA, 95
    scientists of, in Nazi missile programs, 77,
        101
    V-2 rockets of, 79, 101
Gilruth, Robert R., 131
Glasgow, Scotland, 142
Glenn, John, 102
Global Positioning System (GPS), 97, 104, 142
Gobi Desert, 30

Goddard Space Flight Center, 131
Goddard, Robert, 101
GPS. *See* Global Positioning System
Grand Bahama Island, 115
Grand Turk Island, 115
Great Britain, 1
GSLV. *See* Geosynchronous Launch Vehicle
Guiana Space Center, 16, 39–45
 Alcântara as alternative to, 11
Guiyang City, China, 33
Guizhou Province (China), 33
Gulf Coast Regional Spaceport, 148
Gulf of Alaska, 120
Gulf of Mexico, 152

H

H series launch vehicles
 H-I, 59, 66
 H-II, 59, 66
 H-IIA, 59, 63, 66
 H-IIA/H-II, 65
Hawaii, U.S. Western Range instrumentation in, 124
Headquarters 14th Air Force, 125
Houston, Texas, 148
Hudson Bay, 21
human spaceflight, 27
 See also crewed spaceflight

I

ICBM. *See* intercontinental ballistic missiles
Idaho Spaceport, 148
IGY. *See* International Geophysical Year
Il-76 (transport plane), 82
India, 47–50
 space program of, in competition with China's, 27
Indian National Committee for Space Research, 47
Indian Ocean, 110, 121
 in U.S. Eastern Range oversight area, 105
Indian River (Florida), 109
Indian Space Research Organization (ISRO), 47–50
Institute for Atmospheric Physics, 71
Institute of Space and Aeronautical Sciences (ISAS), 57
Institute of Space and Astronautical Science (ISAS), 57
Instituto Nacional de Pesquisas Espaciais (INPE), 9
Intelsat (satellite), 87
intercontinental ballistic missiles (ICBM), 85
 developed by China, 27, 28, 34
 developed by Soviets, 77
 launched at Plesetsk, 79–82

launched at U.S. Western Range, 105, 121, 123, 124
Titan, test flights of, 118
intermediate range ballistic missiles (IRBM), 27, 118, 121
International Geophysical Year (IGY), 21, 57, 77, 101
International Launch Services, 88
International Space Station (ISS), 9, 79, 85, 104
IRBM. *See* intermediate range ballistic missiles
Iridium (satellite), 34
ISA. *See* Israel Space Agency
ISAS. *See* Institute of Space and Aeronautical Sciences; Institute of Space and Astronautical Sciences
Islamabad, Pakistan, 73
Israel, 53–55
 and use of Alcântara launch site, 11
Israel Space Agency (ISA), 53–55
ISRO. *See* Indian Space Research Organization
ISS. *See* International Space Station
Itokawa, Professor Hideo, 57

J

Jacksonville, Florida, 105
Japan, 57–66
 Australian agreement with, for launching experimental planes, 3
 launch of satellites into orbit by, 27
 space program of, in competition with China's, 27
Japan Aerospace Exploration Agency (JAXA), 59, 63, 65
JAXA. *See* Japan Aerospace Exploration Agency
Jiuquan Satellite Launch Center (JSLC), 28–30
 spacecraft too large to launch from, 34
 unsuitability of, for geosynchronous launches, 31
Jiuquan, China, 30
Johnson Space Center, 9
Johnson, President Lyndon B., 102
Joint Defense Facility Nurrungar, 1–3
Joint Long Range Proving Ground, 106
Jonathan Dickinson Missile Tracking Annex, 109, 110, 115
Jonathan Dickinson State Park, 108, 110, 115
JSLC. *See* Jiuquan Satellite Launch Center

K

K-1 reusable launch vehicle, 3, 150, 155
Kagoshima Prefecture, 60
Kagoshima Space Center (Japan), 57, 60
 See also Uchinoura Space Center
Kappa (sounding rocket), 60
Kapustin Yar Cosmodrome, 79

Karachi, Pakistan, 73, 75
Kazakhstan, 78–79, 85–92
Kennedy Space Center (KSC), 104, 105–116
Kennedy, President John F., 77, 102, 105–106
Kern County, California, 149
Khrushchev, Soviet Premier Nikita, 77
Kiku-2 (satellite), 57, 63
Kiribati, 139
Kiritimati, 139
Kiruna, Sweden, 69, 95, 98
Kistler Aerospace Corporation, 3, 150, 155
Kodiak Island, 120
Kodiak Launch Complex, 104–105, 120
Korean War, 117
Korolev, Sergei, 77
Kosmos (launch vehicle), 79
    Kosmos-3M, 81
        launch complex for, at Plesetsk, 82
Kosmos (rocket), 85, 87
Kosmotras, 88
Kourou, French Guiana, 39–41
Krunichev State Research and Production Space
    Center, 88
KSC. *See* Kennedy Space Center
KT-1 (launch vehicle), 36
Kvaerner Group. *See* Aker Kvaerner Group
Kwajalein Atoll, 104
Kwajalein Missile Range. *See* Reagan Test Site
Kyushu, 60, 63

L

Lahore, Pakistan, 73
Lake Hart Air Weapons Range (LHAWR), 4
Lake Michigan, 156
Lambda (sounding rocket), 60
    Lambda 4S-5, 57
Lambda launch complex, 60
Landsat (satellite), 123, 127
Langley Aeronautical Laboratory, 131
Las Cruces, New Mexico, 153
Las Vegas, Nevada, 150
LE-3 (engine), 63
LE-7 (engine), 65
LE-7A (engine), 65
Leeward Islands, 115
Leninsk, Kazakhstan, 85, 87
Leonov, Alexei, 77
LHAWR. *See* Lake Hart Air Weapons Range
Liang Shan, Mount (China), 31
LMLV. *See* Lockheed Martin Launch Vehicle
Lockheed C-5A, 4
Lockheed Launch Vehicle, 120
Lockheed Martin, 120, 124, 148, 150, 154
Lockheed Martin Launch Vehicle (LMLV), 120,
    124
Lockheed Martin Space Operations, SRB re-
    trieval vessels of, 112

Lompoc, California, 120
Long Beach, California, 139, 141, 142
Long March (launch vehicle), 11, 27, 31, 73
    CZ-2E, 28, 33
    CZ-2F, 28
    CZ-3, 33
    CZ-3A, 33
    CZ-3A/B/C family, 33
    CZ-3C, 33
    CZ-3E, 33
    CZ-4B, 34
    Long March (CZ-2C), 28
    Long March (CZ-2D), 28
    Long March 1 (CZ-1), 27
    Long March 2, 34
    Long March 2C/SD, 35
    Long March 2E, 31
    Long March 3, 31
    Long March 4, 34
    Long March 4B, 35
    use of, at TSLC, 35
Los Angeles, California, 118
Luliang Mountains (China), 34
Luna 2, 77
lunar landing, 27, 47, 77–78, 102, 103
Lunar Orbiter 1, 102

M

Madras, India, 49
magnetic field, of Earth
    Chinese-European study of, 34
    launches perpendicular to, from Svalbard, 71
Manitoba, 21
Manned Orbiting Laboratory (MOL), 103, 106,
    114, 118, 123
Mao 1 (satellite), 27
Mars, 47, 104
Mars Pathfinder mission, 104
Mars Polar Lander, 104
Marshall Islands, 104
Martin Marietta, 120
MECB. *See* Missão Espacial Completa Brasileira
Mediterranean Sea, 53, 55
Melbourne Beach, Florida, 112
Merritt Island, Florida, 106
MFTGS. *See* Missile Flight Termination Ground
    System
Miami, Florida, 105
MiG-31 fighters, 87
miniature satellites, Israeli research in, 53
Minuteman program, 127
    booster processing facilities for, 124
Mir space station, 78–79
Mirny, Russia, 79–82
Missão Espacial Completa Brasileira (MECB), 9
Missile Flight Termination Ground System
    (MFTGS), 125

Missile Observation System, 118
Mobile, Alabama, 147
Mojave Spaceport, 149
MOL. *See* Manned Orbiting Laboratory
Molniya
    launch complex at Plesetsk, 82
    launch vehicle, 81, 85, 87
Momo-1 (satellite), 63
Mongolia, overflight of, China avoiding, 30
Montana Spaceport, 150
Moscow, Russia, 81, 87, 145
Mu launch complex, 60, 62
Mu-3SII (launch vehicle), 62
M-V (orbital rocket), 60–62
M-V launch complex, 62

N

N series launch vehicles, 57
    N-I, 57, 63
    N-II, 57, 63
NACA. *See* National Advisory Committee for Aeronautics
Nagatsubo plateau (Japan), 60
NAL. *See* National Aerospace Laboratory of Japan
NASA. *See* National Aeronautics and Space Administration
NASDA. *See* National Space Development Agency
Natal, Brazil
    international airport of, 18
    location of CLBI in relation to, 15
National Advisory Committee for Aeronautics (NACA), 101, 130
National Aeronautics and Space Administration (NASA), 103, 120, 121, 126
    Black Brant launches for, at Woomera, 3
    Clementine Project of, 104
    Earth Observing System of, 122
    Eastern Range facilities of, 112
    formation of, 101
    future role of, 104
    Goddard Space Flight Center of, 15
    inquiries from, about South Dakota Spaceport, 152
    launch of Delta rocket by, 121
    maintenance of FSA tank by, 113
    Manned Spacecraft Center of, 131
    operation of KSC by, 105–108
    radar systems of, at Wallops Island, 109
    Sounding Rocket Program of, 131
    Space Transportation System of, 106, 114
    spacecraft laboratory of, 126
    suborbital launches by, 130–131
    University of Alaska's Geophysical Institute and, 104
    use of ARR by, 69

VentureStar program of, 148
White Sands Ground Station of, 145
National Aerospace Laboratory of Japan (NAL), 3, 59
National Coalition of Spaceport States, 147
National Committee for Space Research (NCSR), Israel, 53
National Engineering and Environmental Laboratory, 148
National Institute for Space Research (Brazil), 9
National Oceanic and Atmospheric Administration, satellites of, 123
National Reconnaissance Office, 122
National Research Council of Canada, 21
National Space Agency, of Ukraine, 141
National Space Development Agency (NASDA), 57–59
National Space Transportation Policy, 108
Navaho weapon systems testing program, 116
Naval Air Warfare Center, 125
Naval Research Laboratory (NRL), 126
NCSR. *See* National Committee for Space Research
Nevada Test Site, 150
New Mexico, 153
Niijima Island, 63
Niijima Test Center, 63
Nike launch vehicles
    Nike-Apache (sounding rocket), 15, 95
    Nike-Cajun (sounding rocket), 21, 69, 73, 95
    Nike-Orion (sounding rocket), 23, 132
    Nike-Tomahawk (sounding rocket), 23, 132
Nike-Zeus Target and Satellite program, 118
Nixon, President Richard M., 102
North Atlantic Ocean, 109
North Korea, 59
Norway, 69–71
    coverage of, by impact area of Esrange, 97
    membership of, in EASP/ESA, 95
    and Sea Launch, 139–141
Norwegian Air Force, 69
Norwegian Defense Research Establishment, 71
Norwegian Sea, 71
Norwegian Space Centre (NSC), 69
NRL. *See* Naval Research Laboratory
NSC. *See* Norwegian Space Centre
Nurrungar tracking station, 1–3
Ny-Alesund, Norway, 69

O

Ofeq military reconnaissance satellites, 53
Ohsumi (satellite), 57
Oklahoma Spaceport, 151
Orbital Sciences Corporation, 123, 127

orbits, 30
  circular, 9
  equatorial, 11
  geostationary, 57, 87, 125
    Chinese launches, 31
    commercial satellites launched into, 41
    equatorial launch location and, 39
    Japanese satellite in, 63
    launching into from Alcântara, 11
  geosynchronous, 147
    and CCAS launches, 108
    Chinese launches, 31
    satellites in, 63
  geosynchronous transfer, 49
  high-inclination, 81, 120
  highly elliptical, 139
  low Earth, 28, 30, 60, 73, 77, 87, 134, 147
  low-inclination, 108
  medium Earth, 28, 139, 147
  polar, 49, 79, 81, 95, 118, 120, 121, 125, 126, 151
    achieved by in-flight azimuth maneuvers, 131
  retrograde, 55
  sun-synchronous
    achieved by in-flight azimuth maneuvers, 131
    Chinese, 34
    Japanese satellite in, 63
Organizing Group for the National Committee on Space Activities (Brazil), 9
Orion (launch vehicle), 23, 132
Orlando, Florida, 108
Osaki Launch Complex, 63, 65–66
Osaki Range, 66
Osaki Rocket Launch Complex, 65
Oslo, Norway, 69

P

Pacific Ocean, 60, 105, 120, 124, 139
PAEC. See Pakistan Atomic Energy Commission
PAFB. See Patrick Air Force Base
Pakistan, 73–75
Pakistan Atomic Energy Commission (PAEC), 73–75
Pakistani National Defence Complex, 75
Palm Bay, Florida, 112
Palmachim Air Force Base, 53
Palmachim Launch Site, Israel, 53–55
Patrick Air Force Base (PAFB), 105–108
Peacekeeper programs, 126
  booster-processing facilities for, 124
Pegasus (launch vehicle), 120, 121
  booster processing, 124
  launches of, at WFF, 130

vehicle assembly building for, 127
People's Republic of China. See China
Persian Gulf War, 3
Pillar Point Air Force Station, 124
Pilotless Aircraft Research Station, 131
Plesetsk Cosmodrome, 79–82
Point Arguello (California), 118, 123
Point Conception (California), 120
Poker Flat Research Range, 104
polar satellite launch capabilities, Chinese, 34
Polar Satellite Launch Vehicle (PSLV), 47–49
Pontes, Marcos, 9
Project Gemini, 106
Project Mercury, 101–103
  spacecraft of, 1, 106, 131
Project Ozone, 15
Prospero satellite (Great Britain), launch of, from Woomera, 1
Proton (launch vehicle), 79, 85, 87
  facilities at Baikonur, 88–91
  Proton K, 88, 89
  Proton M, 88–90
Proton antenna, 145
PSLV. See Polar Satellite Launch Vehicle
Puerto Rico, 115

Q

Quadros, President Jânio (of Brazil), 9

R

R-7 rocket, 77, 79, 85
Ramnan, Mount (Norway), 71
Ramon, Ilan, 53
Reagan Test Site, 104–105
Redstone (rocket), 1
Rehbar (rocket), 73
Remote Launch Control Center (RLCC), at Western Range, 120, 123–124
remote sensing
  Israeli research in, 53
  by Japanese satellite, 63
  satellites for, developed in India, 47
  scientific data from, acquired at SUPARCO, 73
retrograde orbits, 55
reusable launch vehicle (RLV), 147–156
Risoyhamn, Norway, 71
RLCC. See Remote Launch Control Center
RLV. See reusable launch vehicle
rocketry, 1, 77, 101
Rockot (launch vehicle), 82, 85, 87
Rohini-1 (satellite), 47
Rohini-125 (sounding rocket), 49
Rosaviacosmos, 87
Royal Australian Air Force, Aircraft Research and Development Unit of, 3

Royal Norwegian Air Force, 69
RSC Energia, 139, 142, 145
Russia, 77–92, 139–141
    achievement of human spaceflight by, 27
    agreement of SpacePort Canada with, 21
    Brazilian agreement with, for cooperation
        on space projects, 9, 13
    coverage of, by impact area of Esrange, 97
    ground tracking stations, used for Sea
        Launch, 145
    Kvaerner shipyard in, for Sea Launch, 144
    and launch of commercial Israeli satellites,
        55
    missile launches and nuclear detonations
        by, tracking at Nurrungar of, 3
    overflight of, China avoiding, 30
    participation in space race by, 101–103
    Soyuz launch vehicle of, at CSG launch
        site, 41
    technical assistance for Chinese astronauts
        provided by, 27
    Zenit-2 rocket of, 73
Russian Aviation and Space Agency. See Rosa-
        viacosmos
Russian Design Bureau of Transport Machinery,
        82
Russian Federation, 78–81, 85
Rutan, Burt, 147, 149

S

Sakigake (Pioneer) probe, 62
Salyut space station, 78
    Salyut 7, 39, 47
San Diego, California, 117
San Francisco, California, 117
Santa Barbara County, California, 117
Santa Maria, California, 120
Santa Ynez Peak (California), 124, 125
São Luis, Brazil, 11
Sarabhai, Vikram, 49
Satellite Launch Vehicle (SLV), 47
Satish Dhawan Space Center, India, 47–50
Saturn (rocket)
    Saturn 1B, launches of, 114
    Saturn V, 77, 103
    Saturn V/Apollo, 106
        launches of, 114
Scout (launch vehicle), 118, 120, 126, 130, 132
SDO Yuzhnoye/PO Yuzhmash, 139
Sea Launch, 139–145
Seattle, Washington, 154
Shanxi Province (China), 34
SHAR. See Sriharikota Launch Range
Sharma, Rakesh, 47
Shavit (rocket), 11, 53, 55
Shenzhou spacecraft, 27
Shepard, Alan, 102, 106

Shuang Cheng-Tzu, China, 28–30
Shuttle Activation Task Force, 118
Sichuan Province (China), 33
    Xichang region of, 31
Silva, President Luis Inácio da (of Brazil), 9
Sinnamary, French Guiana, 39
Skylab (space station), 103
Skylark (sounding rocket), 95–97
    launch of, from Woomera, 1
    Skylark-12, 15
SLV. See Satellite Launch Vehicle
Snark weapon systems testing program, 116
Sojourner Rover mission, 104
Sonda (sounding rocket), 15
    Sonda-1, 11, 15
    Sonda-2, 11, 15
    Sonda-3, 15
    Sonda-4, 15, 16
Sonmiani Beach, 73
Sonmiani, Pakistan, 73
South America, 15, 39, 108
    Brazil, 9–18
    French Guiana, 39–45
South Atlantic Ocean, 110, 115
South Australia, selection of, for missile systems
        testing, 1
South Dakota Spaceport, 152
South Gannet Hill, 110
South Pacific Ocean, 139
South Texas Spaceport, 152
Southwest Regional Spaceport, 153
Soviet Union, 77–82, 102
    assistance to China provided by, 27
    in Cold War, 1, 101
    collapse of, 101, 139
    French space program and, 39
    launch of satellites into orbit by, 27
    launch of Sputnik by, 77, 101, 106
    and location of Baikonur Cosmodrome, 85
    See also Kazakhstan; Russia; Ukraine;
        U.S.S.R.
Soyuz (launch vehicle), 41, 78, 79, 81, 85, 87
    facilities at Baikonur, 91
    launch complex at Plesetsk, 82
    possible launches of, from Alcântara, 13
Space Activities Commission (of Japan), 57
Space and Upper Atmosphere Research Com-
        mission (SUPARCO) Flight Test
        Range, Pakistan, 73–75
space shuttle, 103
    assembly building, 124
    Challenger, 57, 103, 108, 120, 123
    Columbia, 53, 79, 103–104, 106
    consolidation of operations for, 120
    Discovery, 103–104
    docking with Mir, 79
    emergency landing site for, 154
    flight verification vehicle, 118

halting of program, 103, 104, 118
launch site for, 106
launches of, from KSC, 105, 113–114
to be manufactured by Japan, 59
payloads for, 122
planned Western Range launch facility for, 123
program within U.S. space program, 104
retirement of fleet, 104
Space Transportation System (STS), 103
SpaceLift Australia, 3
SpacePort Canada, 21
Spaceport Florida Authority, 114
Spaceport Sheboygan, 156
Spaceport Washington, 154
SpaceShipOne, 147, 149
SpaceX (corporation), 122
Sputnik
    Sputnik 1, 9, 77–79, 85, 101, 106
    Sputnik 2, 77
Sriharikota Island (India), 49
Sriharikota Launch Range (SHAR), India, 47, 49–50
Sriharikota Space Centre, India, 49–50
SSC. *See* Swedish Space Corporation
Starsem, 87, 91
Start (launch vehicle), 21, 81, 82
Stavanger, Norway, 144
Straits of Gibraltar, 55
STS. *See* Space Transportation System
Sudden Ranch, 118
Suisei (Comet) probe, 62
SUPARCO. *See* Space and Upper Atmosphere Research Commission
Super Loki (sounding rocket), 15, 16, 132
Surveyor 1, 102
Svalbard, archipelago of, 69–71
Svobodny Cosmodrome, 79
Sweden, 95–99
    coverage of, by impact area of Esrange, 97
    ESRO-sponsored launches in, 69
    membership of, in EASP/ESA, 95
Swedish Space Corporation (SSC), 95
Switzerland, membership of, in EASP/ESA, 95

T

Taihang Mountains (China), 34
Taiyuan City, China, 34
Taiyuan Satellite Launch Center (TSLC), 34–36
Takesaki Launch Complex, 63
Takesaki Launch Site, 66
Tanegashima Island (Japan), 63, 65
Tanegashima Space Center (TNSC), 57–59, 63–66
Taurus (launch vehicle), 120, 121, 123
    booster processing, 124

Taurus-Orion (sounding rocket), 23
TCLC. *See* Titan Centaur Launch Complex
TDRSS. *See* Tracking and Data Relay Satellite System
Tel Aviv, Israel, 53
Telstar 1 (satellite), 102
Tequesta, Florida, 110, 115
Tereshkova, Valentina, 77
Terra (satellite), 122
Texas, 104, 148, 152, 155
The Aerospace Corporation, RLCC study by, 123
The Boeing Company, 114, 120, 139
    Delta IV launch system of, 121
    transformation of SLC-6 by, 123
30th Communications Squadron, 124
30th Weather Squadron, 125
Thor (IRBM), 118, 120, 121, 122
    launch of, 113
    Thor/Agena (booster combination), 118
    Thor-Delta, 63
Thumba, India, 47, 49
Tiamat (model rocket), 131
Tiros 1 (satellite), 102
Titan (ICBM), 118
Titan (launch vehicle), 108, 120
    Titan 34D, 114, 123
    Titan I, 113
    Titan II, 113, 123, 124
        in Gemini launch, 106
    Titan III, 122
    Titan IIIA, 113
    Titan IIIB, 118
    Titan IIIC, 114
    Titan IIIM, 118
    Titan IV, 108, 114, 121, 122, 124
    Titan IVA, 122
    Titan IVB, 114, 122
Titan Centaur Launch Complex (TCLC), 123
Titan, launchpads for, 82
TNSC. *See* Tanegashima Space Center
Tokyo, Japan, 57, 63
Tomahawk. *See under* Nike
TR series rockets
    TR-1, 66
    TR-1A, 66
Tracking and Data Relay Satellite System (TDRSS), 145
Tranquillion Peak (California), 125
Trivandrum, India, 49
TSLC. *See* Taiyuan Satellite Launch Center
Tsyklon (launch vehicle), 82, 85, 87
    Tsyklon-3, 81
    Tsyklon-4, 41
TU-154 (transport plane), 82
Tyuratam, Kazakhstan, 85

U

U.S. Air Force, 21, 103, 114, 116, 118–120, 123,
    126
    announcement by, regarding SLC-6 con-
        struction, 118
    Atlas launchpads for, 122
    Camp Cooke turned over to, 117
    Cape Canaveral Air Station of, 105
    Dyna-Soar program of, 101, 103
    45th Space Wing of, 105
    4950th Test Wing of, 110
    initiation of Atlas/Centaur development
        by, 122
    IRBM program of, 121
    meteorological support for, 112
    MOL program of, 114, 123
    reliance on ELVs by, 106
    shuttle launches from Western Range,
        preparation for, 118
    Space Command of, 1, 34
    testing programs of, 121
    30th Space Wing of, 117
        missile flight control requirements
            for, 125
        mission of, 120
        operation of optical systems by, 125
        payload processing by, 125
    use of Titan launch vehicle by, 106
U.S. Army, 21, 106, 117
U.S. Defense Support Program (DSP), 1–3
U.S. Department of Defense (DOD), 121, 122,
    124, 131
    Clementine Project of, 104
    space program of, 101–103
U.S. Department of Energy (DOE), 149, 151
U.S. Marine Corps, 149
U.S. Missile Defense Agency, 105
U.S. Naval Facility Argentia, 115
U.S. Navy, 106, 110, 115, 118, 122
    operation of Ascension Auxiliary Air Field
        by, 115
    Scout launch vehicles of, 126
    Trident II ballistic missile of, 114
    Vandal missile of, 132
U.S.A. *See* United States
U.S.S.R., 63
Uchinoura Space Center (USC), 57, 59, 60–62
Ukraine, 78, 88, 92, 139, 141
    and use of Alcântara launch site, 11
United Kingdom, Australian agreement with, for
    weapons systems development, 1
United States, 63, 101–137, 139
    achievement of human spaceflight by, 27
    agreement with Japan to import technolo-
        gy, 57
    Brazilian agreement with, for astronaut
        training, 9
    in Cold War, 1
    competition by, against Soviet Union, 77
    French space program and, 39
    launch of satellites into orbit by, 27
    proposed commercial spaceports of, 147–
        156
University of Alaska, Geophysical Institute of,
    104
University of Rostock (Germany), 71
University of Tokyo, 57
USC. *See* Uchinoura Space Center
Utah Spaceport, 154

V

V-2 rocket program, 101
VAFB. *See* Vandenberg Air Force Base
Van Allen radiation belts, 102
Vandal (missile), 132
Vandenberg Air Force Base (VAFB), 104, 105,
    117–128
Vandenberg Telemetry Receiving Site, 124
Vandenberg, Gen. Hoyt S., 118
VCSFA. *See* Virginia Commercial Space Flight
    Authority
Vega (launcher), 41, 42
Veículo Lançador de Satelites (VLS) rocket, 9–
    11
VentureStar, 148, 150, 151, 154, 155
Véronique (sounding rocket), 39, 41
Virgin Islands, 115
Virginia Commercial Space Flight Authority
    (VCSFA), 132
Virginia, commercial launch capability for, 132
Volgograd Station. *See* Kapustin Yar Cosmo-
    drome
Volgograd, Russia, 79
von Braun, Wernher, 101
Vostok (spacecraft), 78
    Vostok-2, 81
    Vostok-6, 77
Vyborg, Russia, 144

W

Wallops Flight Center, 131
Wallops Flight Facility, 104, 130–136
Wallops Island, Virginia, 130–136
    radar systems at, 109
Wallops Orbital Tracking Station (WOTS), 134
Wallops Station, 131
WCOOA. *See* West Coast Offshore Operating
    Area
West Coast Offshore Operating Area
    (WCOOA), 120
West Texas Spaceport, 155
Western Range (United States), 104, 117–128
Western Range Control Center (WRCC), 124

Western Test Range, 117
WFF. *See* Wallops Flight Facility
WIR. *See* Woomera Instrumented Range
Wisconsin Spaceport, 156
Woomera Instrumented Range (WIR), 1, 3–4
    launch areas within, 4
Woomera Prohibited Area (WPA), 1–7
Woomera Rangehead, 3
Woomera Rocket Range, Australia, 1–7, 150
    See also Woomera Prohibited Area
Woomera township, 1, 4
World War II, 57
    air operations during, 116
    French space program and, 39
    German scientists captured during, 101
    Soviet rocketry program and, 77
    U.S. entry into, 117
WOTS. *See* Wallops Orbital Tracking Station
WPA. *See* Woomera Prohibited Area
WRCC. *See* Western Range Control Center
Wuzhai Missile and Space Test Center. *See*
    Taiyuan Satellite Launch Center

X

X-33, 148, 150, 155
Xichang Satellite Launch Center (XSLC), 31–34
    as model for TSLC, 35
Xichang, China, 31, 33
XSLC. *See* Xichang Satellite Launch Center

Y

Yang Liwei, 27
Yangel Yuzhnoye Design Bureau, 88
Yavne, Israel, 53
Yibin, China, 33
Yoshinobu Launch Complex (Japan), 63–65, 66

Z

Zenit (launch vehicle), 81, 85, 87
    facilities at Baikonur, 91
    Zenit-2, 73
    Zenit-3SL/Block DM-SL, 139, 145
Zhongxing-22 (satellite), 31
ZY-2C (satellite), 34